CONTENTS

VNIITE
DISCOVERING UTOPIA:
LOST ARCHIVES
OF SOVIET DESIGN

ALEXANDRA SANKOVA
OLGA DRUZHININA

This publication is dedicated to the VNIITE's
employees – Soviet designers who remained
dedicated and passionate professionals,
despite the challenges they faced, and who
became our good friends and the best
kind of ambassadors for the Moscow Design
Museum. For them, design was not just
their profession: it was their life.

VNIITE
DISCOVERING UT
LOST ARCHIVES
OF SOVIET DESIG

ALEXANDRA SANF
OLGA DRUZHININ,

This publication is dedicated to
employees – Soviet designers v
dedicated and passionate profe
despite the challenges they fac
became our good friends and th
kind of ambassadors for the Mo:
Museum. For them, design was
their profession: it was their life.

3

Writing in Design magazine in 1967, Frank Height, the Royal College of Art's Professor of Industrial Design, reported back from a visit to the USSR and a tour of its All-Union Scientific Research Institute of Technical Aesthetics (VNIITE), five years on from its founding. By this point "the largest single industrial design organisation in the world" had established its Moscow headquarters and a network of offices in Leningrad, Kharkov, Sverdlovsk, Khabarovsk, Minsk, Kiev, Vilnius, Tbilisi and Yerevan.

Height was impressed by what he saw and was eager to see where the VNIITE would take Soviet design. He praised its "knowledgeable and enthusiastic designers, supported by specialists in the techniques of modern industrial life and with powerful information services". The VNIITE "serves an enormous and varied industrial complex," Height wrote, "and because of its official position can speak to industry with an authority many Western designers might envy"[1].

Some fifty years on from these observations, it is hard to fathom how such an institution, dedicated to the promotion of design, in theory and in practice, the improvement of design standards within the Soviet Union and its international reputation could have faded so far from view. There are scant references to the VNIITE in Western design histories, while in Russia, the history itself has been gradually slipping away, largely unrecorded. Thanks to the efforts of the Moscow Design Museum, however – and the personal archives of some of the VNIITE designers – the story of this remarkable organisation is being pieced back together. It is a design heritage that is slowly emerging from obscurity.

In retelling the story of the VNIITE, the images in this book act as a visual record of its approaches, programmes, seminars and publications (such as its monthly journal Technical Aesthetics) and reveal an organisation striving for a better Soviet society. Few of the VNIITE's products were ever made, let alone mass produced, but in its array of concept drawings, models and prototypes we see glimpses of how life in the USSR might have been improved by design – and echoes of where design movements outside of Russia were already heading. The institute was always able to look beyond the Soviet Union (its Director, Yuri Soloviev was committed to forging international relations) and in fact the VNIITE became one of the most outward-facing components of the USSR state system.

Ironically, it was in part the inability to forge the important links with industry that Height had noted in the late 1960s that would ultimately lead to the failure of the VNIITE. In 1999, Dmitry Azrikan, one of the institute's most well known designers looked back on its achievements and its struggles and wrote that a "unique design philosophy, based on profound humane values can be found in the buried VNIITE heritage. It should be unearthed and studied."[2] Now, the rediscovery of the VNIITE should be welcomed just as its original efforts were half a century ago.

Mark Sinclair,
Senior Editor, Unit Editions

1.   'Design in the Soviet Union' by Frank Height, Design 228 (December 1967)

2.   'VNIITE, Dinosaur of Totalitarianism or Plato's Academy of Design?' by Dmitry Azrikan, Design Issues, vol. 15, no. 3 (Autumn, 1999), The MIT Press

DESIGN IN THE USSR AND THE BACKGROUND TO THE VNIITE

ALEXANDRA
SANKOVA
&
OLGA
DRUZHININA

The VNIITE offices were based in a pavilion in Moscow's Exhibition of Achievements of the National Economy

Attendees at the IX ICSID Congress which took place in the Rossiya hotel in Moscow, 1975

Yuri Soloviev, Director of the VNIITE

The All-Union Scientific Research Institute of Technical Aesthetics (VNIITE) was a unique creative platform. 'Vniitians', as the organisation's employees called themselves, designed for the future and developed new theories and approaches to design. The atmosphere at the institution was particularly conducive to creativity: the most progressive of Moscow's leading art experts and designers worked there. Yet to understand how and why such an institute was established in the early 1960s, we need to explore the development of Soviet design in the first half of the 20th-century.

Design emerged as a profession in the USSR in the 1920s, when leading figures of the Russian avant-garde moved towards creating art for social purposes. The Constructivists[1] founded the first schools of design for furniture and products, ceramics and textiles within the industrial faculties of the Higher Art and Technical Studios (VKhUTEMAS). These institutions were established by a decree of the Council of People's Commissars signed by Vladimir Ulyanov (Lenin) for the purpose of training artists for industry and developing industrial art.

The conceptual centre of the development of Soviet design in the 1920s was the Institute for Artistic Culture (INKhUK). It was here that Alexei Gan, Alexander Rodchenko, Varvara Stepanova, the brothers Vladimir and Georgii Stenberg, Konstantin Medunetsky and Kārlis Johansons formed the First Working Group of Constructivists. In his pamphlet, Constructivism (Tver, 1922), Gan set out this new form of design activity:

"Constructivism differs from the styles of the past in that it is not a stylisation and does not have 'decor'; the form of an object derives from the practical use of a material and its purpose, and the method of construction lies at the basis of the creation of things". The Constructivists were 'industrial artists' and essentially the first Soviet designers, but few of their ideas were put into practice. Ninety percent of their designs remained on paper.

This was nothing unusual for Soviet design. In 1917, Russia, worn out by the First World War, famine and political crisis, was swept by revolution. A civil war began. Companies and factories were nationalised and the country's industry collapsed. While some very small-scale production of textile products and ceramics took place, design in the 1920s was at its most powerful and eye-catching in graphic design – evident in periodicals, books and posters – and in the work created in support of events and Soviet pavilions at exhibitions. In short, propaganda – one of the most important tools for establishing the state ideology.

The 1930s saw the establishment of a small number of projects that involved designers: the construction of the Moscow Metro and the development of the first all-metal ANT-type passenger aircraft (the acronym referred to the initials of the pioneering Soviet aircraft designer, Andrei Nikolayevich Tupolev) and the giant 'propaganda plane', Maxim Gorky. The first aerodynamic steam, diesel and electric locomotives also appeared at this time. However, these were for the most part isolated developments where the role of designer was often played by an engineer, reflecting the thinking that the key factors determining the appearance of an object were its structure and function. Despite famine and huge financial problems, the new state was becoming increasingly industrialised, putting enormous effort into constructing new cities, coal-fired and hydroelectric power stations, factories, canals and bridges.

Gradually, stability returned, but in 1941 the USSR entered the Second World War and design had to take a back seat for several years. The country's urgent need to develop new military hardware meant focusing on the production of equipment – the T-34 tank, the Ilyushin Il-2 ground-attack aircraft, the Shpagin PPSh submachine gun, for example – and so, in the immediate post-war years, the defence industry would serve as a foundation for the development of civilian industry. In 1946, a year after the end of the war, the Architecture and Art Bureau (AKhKB) was set up as part of the Ministry of Transport Engineering in order to improve passenger transport in the USSR. Founded by Yuri Soloviev, then a relatively young graduate of the Moscow Polygraphic Institute, the Moscow Architectural Institute and the Diplomatic Academy, the AKhKB established designs for a range of projects, from space-saving furniture to trolleybuses and passenger ships.

7

We met Soloviev in 2012, when he was 92, and he explained how, based on his own experience, he believed that none of these institutions had provided him with training in 'industrial design', the field in which he had wanted to work. However, the broader education that he received was to prove useful in developing the skills required to take on complex industrial projects involving specialists from different fields and different countries. Soloviev realised that the USSR needed to study and learn from foreign design practices. Once the AKhKB was up and running, he went through several teams of employees – as the country's colleges and universities were not producing industrial designers, people from other professions (typically architects and engineers) had to be retrained.

Soloviev's friendship with Vasily Stalin (the son of Joseph Stalin) and high-ranking party officials helped him to win his first contracts. He was the son of a military engineer, who went on to head up an aircraft factory, and many of his father's friends (and fellow students who had lived in the same student accommodation) also went on to occupy important positions in both the defence and civilian industries. While Soloviev received his first order through acquaintances of his father, this was no guarantee of success, however.

Indeed, due to the way the system worked, such patronage only helped to ensure that designs could be entered into competitions. And following the war, a hugely significant one was announced. A vast number of people were moving across the USSR, as schools, universities and colleges, museums and residents of entire cities returned from evacuation, but the country's rail carriages were totally ill-equipped (they were even called 'cattle wagons' as they were similar to the ones used to transport livestock). Realising that new carriages were required, the government launched a competition to propose a redesign – and Soloviev's team took part. Here, he showed his talent not just for industrial design, but also for organising and marketing.

While the other teams presented passenger carriages with conventional rigid angled seats, Soloviev's group designed them to be folding and ergonomic. The ministry representatives were finally convinced by the mock-up that his team produced, which was on a 1:1 scale – an unheard-of extravagance at the time. Sitting in the 'coupé' (a sleeper carriage divided into compartments) and 'platzkart' (an open-plan sleeper carriage, from the German 'Platzkarte'), the commission decided to give the go-ahead to the comfortable cars. The AKhKB's first project was an all-metal platzkart carriage.

Following this, the bureau developed new passenger rail carriages, postal 'aerosledges', trolleybuses, even types of passenger ship. Effectively the forerunner of the VNIITE, the AKhKB became the country's first specialised art and design bureau responsible for industrial design in the USSR. Yet by the end of the 1950s it was clear that the amount of design work needed to rebuild the country and restore its post-war industry was too much for one organisation to cope with.

Two important events then took place: the 20th Congress of the Communist Party of the Soviet Union in 1956 and the World Festival of Youth and Students a year later. At the Communist Party Congress, Nikita Khrushchev denounced Stalin's cult of personality[2] and what was to become known as the 'Khrushchev Thaw' began. The Soviet Union's borders opened up a little and the first international exchanges of experts took place in Moscow. In 1957, Moscow hosted the sixth World Festival of Youth and Students, giving residents of the capital the chance to encounter art and culture from other countries. Despite their opposing economic systems (and the competition at the heart of the 'space race'), the USSR and the USA signed their first scientific and cultural exchange agreement in 1958. The first events under this agreement were the Soviet National Exhibition in New York City and American National Exhibition in Moscow's Sokolniki Park, both held in 1959.

The Soviets saw works by the American designers Charles and Ray Eames, Richard Buckminster Fuller and George Nelson, the latest in American-made automobiles, modern kitchens, home appliances and computing. The 'Kitchen Debate', a series of exchanges at the exhibition between Khrushchev and US Vice President Richard Nixon, was broadcast to the nation on television. While the former spoke with pride of the USSR's achievements in space exploration, the latter pointed to the fact that every American housewife had a comfortable kitchen.

Nevertheless, the Soviet Union decided that it needed to catch up with the capitalists – and potentially overtake them – in the production of consumer goods (not to mention that the Soviet people were tired of the constant shortages). On the other hand, the quality of life was improving, and there were regular exhibitions in Moscow demonstrating the latest goods and experimental products being made by new enterprises and factories, alongside a growth in new projects from research institutes and by students from specialist higher education institutions. The production and range of consumer goods needed to be increased.

To address these problems, in the late 1950s design departments were set up at colleges of art and industry in Moscow and Leningrad and, in 1961, a division of the Soviet State Research Coordination Committee was set up to organise design activities and to develop relevant measures. Its proposals were approved by the government and incorporated into Resolution 394 of the Soviet Council of Ministers, 'On improving the quality of engineering products and cultural and household goods through the implementation of methods of artistic design', of 28 April, 1962.

That same year, the All-Union Scientific Research Institute of Technical Aesthetics (VNIITE) was established in Moscow, with Soloviev as its director. By this time, he had gained vast experience of designing and implementing large-scale projects and he knew that only centralised management, education and product quality control could improve the standard of design in the USSR. The VNIITE became part of the State Committee for Science and Technology and a whole network of research institutes and specialist design bureaus, each linked to a particular ministry, was established. This included the Experimental Research Institute of the Toy; the All-Union Institute of Furniture Design and Technology (VPKTIM); and the All-Union Institute for Light Industry Products (VIALEGPROM), while design services at various optical and automobile factories were also established[3].

Generally the research institutes had several functions within their sector: centrally coordinating activities, providing training and development, developing future products, staging exhibitions of Soviet and Western design, collaborating with equivalent institutions abroad, obtaining and sharing information, and developing standards. In addition they published trade journals, promoting design as a part of the development process.

From the late 1950s, the Soviet Ministry of Trade and Chamber of Trade and Industry had established 'sample product rooms' where Western consumer goods could be purchased. Manufacturing employees could come and choose a consumer product they liked (an iron, vacuum cleaner, coffee maker, television, and so on), bring it back to their factory, take it apart, study it, and copy it. The resulting products were often far removed from the originals, as the USSR had access to different materials and technical capabilities compared to those countries where the samples had been obtained. Because of the lack of professional design training in the post-war years, and the fact that Russian industry was not advanced enough for the country to manufacture its own products, plagiarism made sense.

Yet in the mid-1960s, Western products were being copied at the same time as original design work was being carried out at the VNIITE. In fact, the existence of the sample product rooms greatly impeded the institute's work and did little to help the development of industrial design in the USSR. There were numerous cases of promising Soviet designs, worked on for anything from three to seven years, being shelved in favour of copies of Western equivalents. Indeed, prizes were even awarded to those adopting Western products and reproducing them with the most accuracy.

Throughout the 1960s and 70s, at the same time as the VNIITE's central Moscow office was growing, ten regional branches were opened up in major USSR republics. Additionally, a branch division network of specialised construction design offices (CDOs) was set up, as well as other industry-specific design services that received recommendations on particular methods to be used from the central office. The design services were deeply involved with local industry-specific projects, as well as optimising the way designers communicated with engineering specialists and other industrial workers.

VNIITE subdivisions were focused on pilot production, where designers made scale models and samples of goods; ergonomics; expert reviews of industrial goods; and on the standardisation of decorative and finishing materials. In 1964, the VNIITE had launched its own specialist journal called Technical Aesthetics and it became the main source of information in the USSR on Soviet and international design. The institute also introduced a government certification for industrial goods in the form of a 'Sign of Quality' in 1967, as well as expert reviews to determine grades for engineering designs before they were mass produced.

The VNIITE also actively worked with other countries. In 1964, at the Exhibition of Achievements of the National Economy (VDNKh) in Moscow, a British design exhibition was staged entitled 'The Role of the Industrial Designer in British Industry', organised by the VNIITE and the British Design Council. In subsequent years, up until the 1990s, the VNIITE regularly held foreign design exhibitions in the USSR – and domestic design exhibitions abroad. In 1965, the VNIITE became a member of the International Council of Societies of Industrial Design (ICSID). (In 1977, Soloviev was voted its President, the only time the organisation was led by a designer from the Soviet Union.)

As with other research institutes, the VNIITE often had to fight against the circumstances in which it operated. On the other hand, such institutes could set trends in Soviet design. Their designers were able to travel to other countries and to international conferences and exhibitions; they were familiar with global trends, not least because they had access to Western trade publications (including Design from the UK). Design was well-promoted within the country and there was cooperation with foreign professionals in sharing best practices.

These efforts, however, were largely academic: designers continued to lead separate professional lives from the manufacturing industry. While many worked in close collaboration with factories and their technology departments (if the designs were intended for production), at other times designers would come up with promising concepts independently (these were usually futuristic or speculative, far from the reality of actual available technologies and/or materials). Yet these designs should have pointed the way for industry and given it new targets.

In practice there was a significant gap between industry and design in the USSR. Research institutes specialising in design attempted to bridge it but were constrained by the realities of their situation. Designers faced huge challenges in getting their projects adopted. A tiny proportion of designs went into production but many promising projects failed to get off the ground and, despite the designers' input, the factory products differed greatly from their original design concepts.

After the collapse of the Soviet Union, the VNIITE suffered a sudden loss of status and was no longer responsible for coordinating and overseeing the quality of design. Many VNIITE designers emigrated. Holistic and systematic, their approach to design was independent of nationally- and technologically-specific features of production. This enabled them to become successful in the West. After 1990, only the VNIITE's Moscow office remained, and much of its material was lost. The private archives of the designers working at the VNIITE were transferred to the Moscow Design Museum – the current publication is based on this material.

In 2014, the VNIITE was absorbed into the structure of the Moscow State Technological University/Moscow Institute of Radio Engineering, Electronics and Automation. The following year, the state-run company Rostec founded the Centre of the All-Russian Scientific Research Institute for Industrial Design Aesthetics, which continues the VNIITE's tradition of work in the field of industrial design.

1. Constructivism was a movement in 1920s Soviet art and it proved to be influential on architecture, product design, typography, set design, posters, book art and literature. Its proponents aimed to 'construct' an environment that actively directed living and sought to understand the potential of modern technology and how its logical, practical design approach could influence form as well as the aesthetic qualities of materials such as metal, wood and glass. In its first years (1918—1921), Constructivism was closely connected with 'leftist' trends in painting and sculpture. As a movement it set out to blend art, which, with the growth of capitalism had become separate from craft, with material production, based on advanced industrial engineering. The Constructivists believed that new architecture and new kinds of residential and public buildings, furniture, equipment and clothing to be the most important means of the Socialist reconstruction of society. They contrasted the showy extravagance of bourgeois life with the simplicity and utility found in the new object forms, which they believed embodied democratic values and the new relationships between people.

2. The 20th Congress of the Communist Party of the Soviet Union took place in Moscow on 14-25 February, 1956, eight months early due to the need to review the changes that had taken place since Stalin's death and to discuss the way forward. The Congress culminated with Nikita Khrushchev's famous 'secret speech' where, on the final day of the congress, the Party's First Secretary presented a report to a closed session entitled 'On the Cult of Personality and Its Consequences' which condemned Stalin's use of mass terror in the Great Purge.

3. Examples include the S.A. Zverev Krasnogorsk Optical and Mechanical Factory (Zenit), the I.A. Likhachov Automobile Plant (ZIL), the Volga Automobile Plant (VAZ) and the V.I. Lenin Leningrad Optical and Mechanical Association (LOMO) factories.

In the early 1960s, design in the USSR became part of state policy. The All-Union Scientific Research Institute of Technical Aesthetics (VNIITE) was created in 1962, with Yuri Soloviev as its Director (pictured on pages 6 and 14), and it became the leading organisation in the field of design in the USSR. Its remit was vast: it developed and implemented artistic design methods, approved scientific research work in the field of industrial design aesthetics, defined design requirements for goods and was responsible for international collaborations. The main office (the grand exterior of which is shown on page 6), was located in Moscow's Exhibition of Achievements of the National Economy (VDNKh), a vast park and exhibitions centre in the Ostankinsky District of the city. The institute also had ten branches operating in major cities and Soviet republics from the Baltics to the Far East. In the 1970s, the first Centre for Industrial Design Aesthetics in the Soviet Union, a special subdivision set up to promote the implementation of design methods within domestic industry, was established in the Izvestia Publishing House building on the capital's Pushkin Square.

VNIITE designers present a model of a building
to the institute's director, Yuri Soloviev

VNIITE design meeting and three designers with
a selection of products

Meeting at one of the Construction
Design Offices

Designers at work in a CDO (and over)

Specialised construction design offices (CDOs) played an important role in the Soviet state system of design. The CDOs worked closely with the design offices at various organisations and scientific research institutes, participated in solving local industry challenges and collaborated with engineers and other specialists involved in the manufacturing process. The largest construction design offices were at the Ministry of Light Industry and Food Industry (CDO LEGMASH) and at the Ministry of Electronic Industry (CDO Estel). The CDO LEGMASH was set up in 1962 and was initially involved in almost every field of manufacturing from heavy industry to garment production. The CDO Estel, created in 1974, bore responsibility for more than 500 organisations that manufactured production machinery, consumer electronics and specialised electronic equipment. The industrial design services in the optical and mechanical industries, such as those at the Krasnogorsk Optical and Mechanical Plant and the Leningrad and Belorussian Optical and Mechanical Unions (LOMO and BelOMO), were also active at this time.

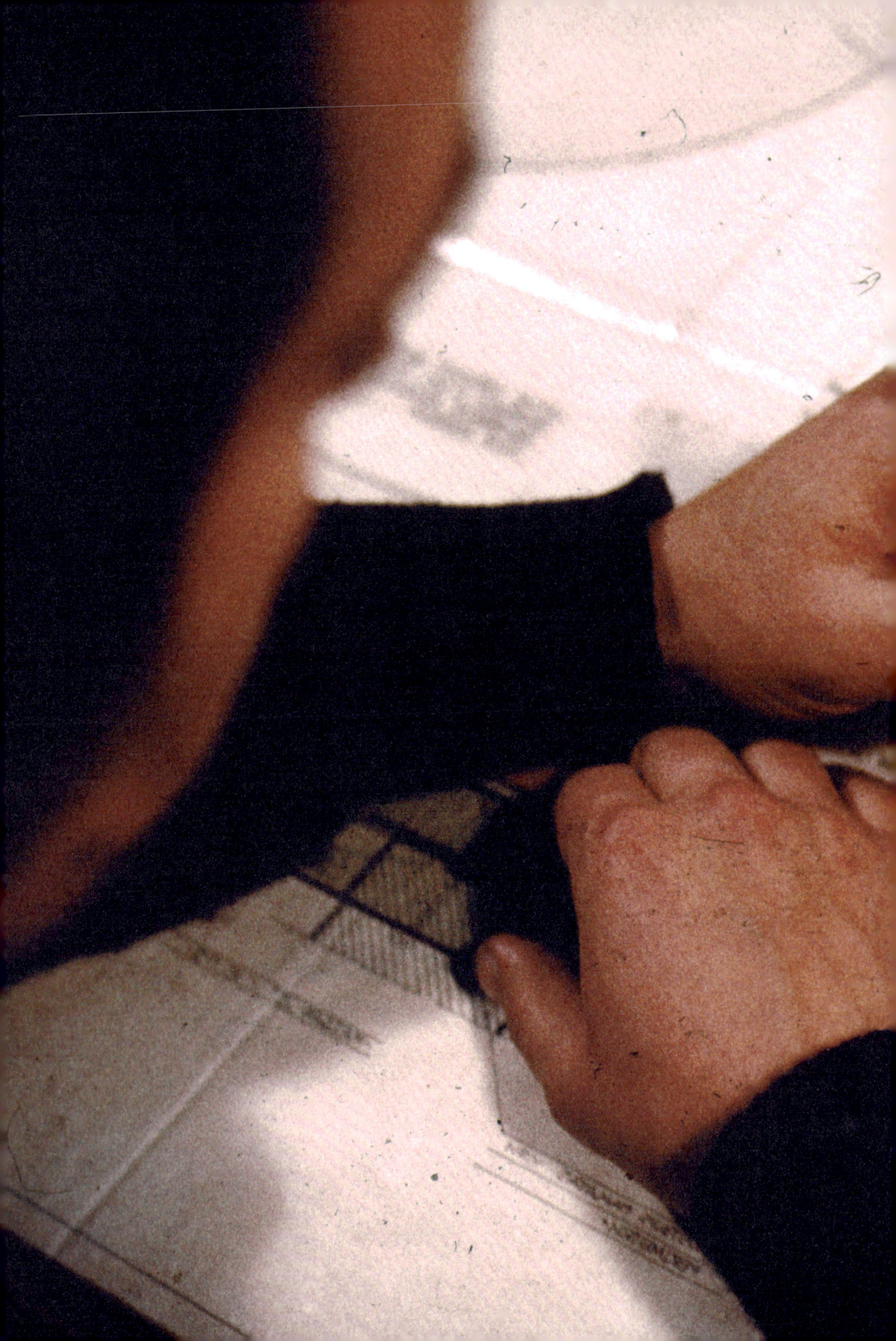

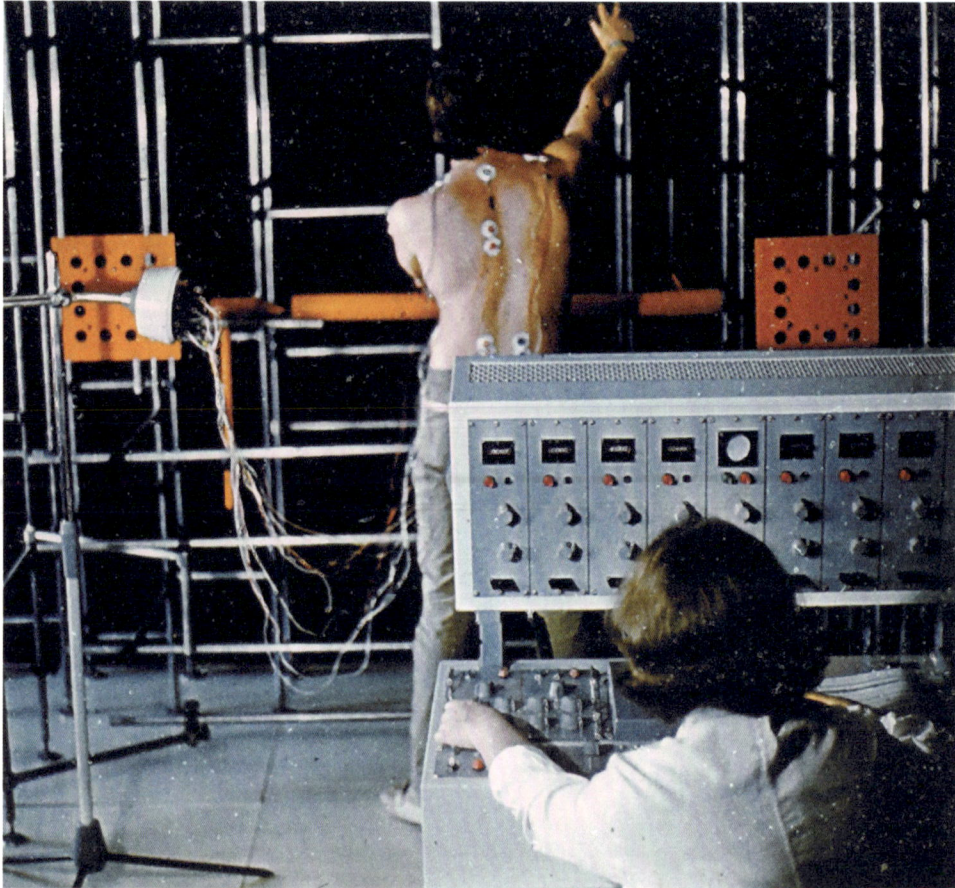

Testing in the VNIITE's ergonomics department

The VNIITE department of ergonomics studied the interaction between humans, machines and the environment. The field of its research included examining different types and forms of human activity (namely its anthropometric, physiological, and psychological aspects) and developing requirements for new equipment in the field of ergonomic design. From the 1960s, the VNIITE developed a national school of ergonomics and created the basis for international scientific and technical collaborations in this field. In 1976, a coordination centre for the Council for Mutual Economic Assistance (COMECON) member countries was organised at the VNIITE in order to spotlight ergonomic issues and was led by the academic Vladimir Munipov, one of the organisation's founders. One particular area that COMECON member representatives examined was the formulation of scientific grounds for developing ergonomic norms and requirements. Following the project a landmark work 'Ergonomics: Principles and Recommendations' was published, founded on data from anthropometric studies.

Testing in the VNIITE's ergonomics department

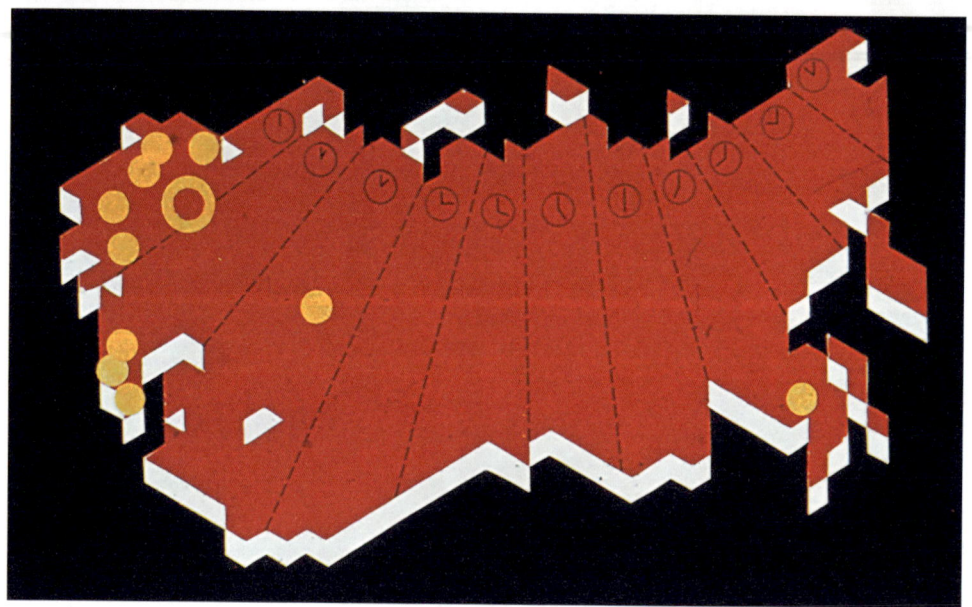

Inside one of the VNIITE subdivisions

Map of the USSR highlighting the VNIITE Moscow
headquarters and its ten regional offices

The VNIITE had ten regional subdivisions located in the USSR's major industrial centres. Generally speaking, the specialist knowledge of a branch office would be linked to the main industry found in their location. Branch employees were engaged in scientific research on the issues of industrial design aesthetics and ergonomics, project work in collaboration with regional industries, developing manufactured goods and design programmes, graphics and packaging, environmental design projects, and providing advisory support to design services. All of the branches collaborated closely and were managed by the central office in Moscow. VNIITE branches also participated in certifying goods for the government-approved Sign of Quality system (see page 48) and all subdivisions took part in major scientific research programmes and developments initiated by the central institute. The leadership at the VNIITE held nationwide and international conferences at its regional subdivisions in order to help employees further their professional development.

Testing a model for a vehicle in the VNIITE
Belarus regional office

Regional VNIITE team

Over: Library collection of design magazines,
including Design from the UK, at one of the
VNIITE's offices

Model wheelchair and, over, model laundry
machines

Modelmaker at work at the VNIITE pilot
production facility

In 1964, a VNIITE pilot production facility was opened in the VDNKh area of Moscow. Here, specialists made scale models of various goods from radio components and milling machines, to cars, including the model of the famous 'Next-Generation Taxi'. The prototyping machines enabled the creation of fairly complicated construction designs. Work on a prototype began by constructing a wooden carcass upon which a model could be fashioned from putty – it was from this form that a scale model could then be built from fibreglass. Working prototypes were constructed at full size and automobile models were even capable of movement. Later, the designers began to use a vacuum-forming process, creating scale models out of plastic. A small group of design engineers who worked at the pilot production facility also ran the paint shop which determined the correct paint compositions for the model in question and studied how the colour and form of the object interacted. The VNIITE also set up a production facility in the city of Mukachevo in the Ukrainian Soviet Socialist Republic.

Modelmakers at work at the VNIITE pilot
production facility

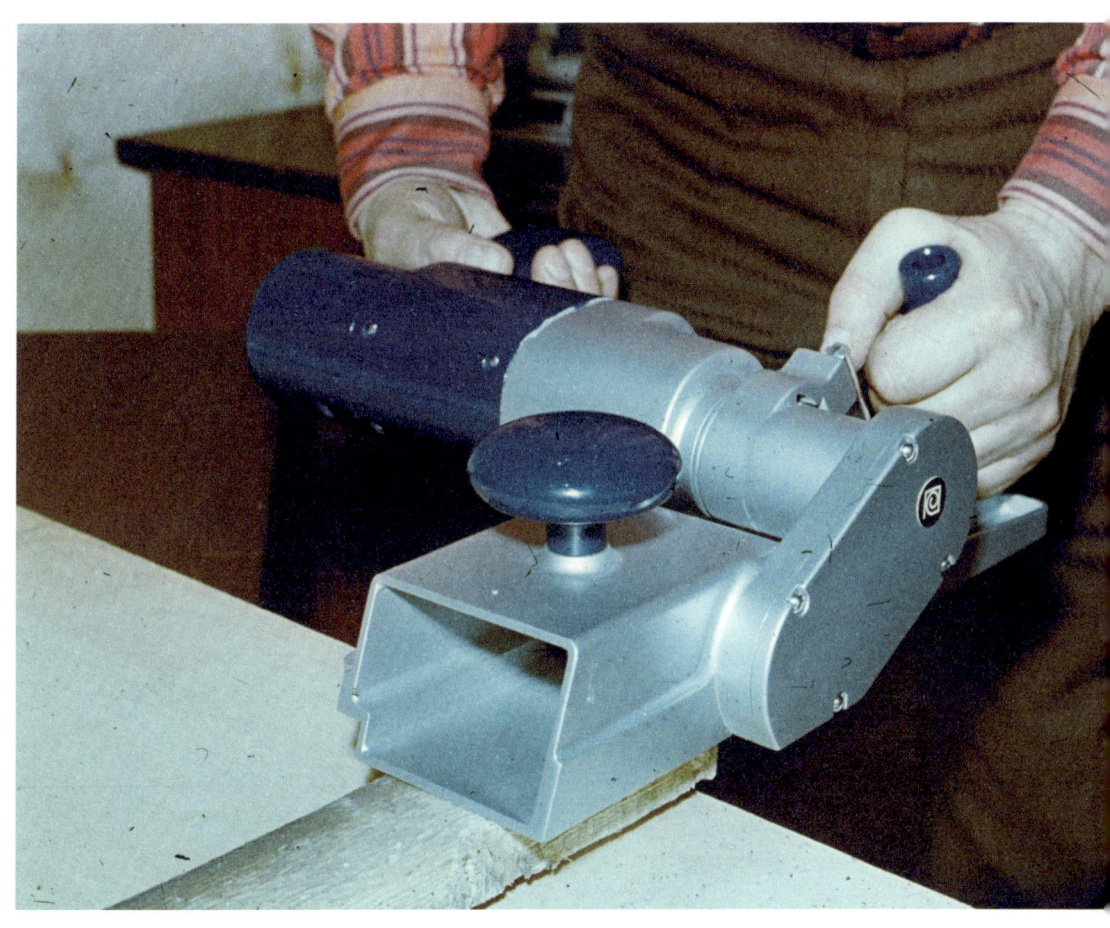

The TDT-55 tractor, designed for hauling timber

The TDT-55 crawler tractor was designed for hauling and stacking timber. This low-maintenance vehicle was adapted to work (and be repaired) in any conditions. Designers created its unique sturdiness by using a range of new, durable materials and rejecting unnecessary mechanisms that regularly malfunctioned. The freed up space in the vehicle was occupied by an improved gearbox and powerful engine which enabled an increase in the model's terrain-crossing ability by 30-40%. The TDT-55 was recognised as a leader in the field of timber processing, but was not limited to the timber industry – its main job was transporting, rolling and positioning heavy loads and weights. The TDT-55 received a number of prizes and medals at international exhibitions, including awards for the industrial design work behind its reliable and comfortable cab. The model was created in 1964 by the designer Tatiana Samoilova, in collaboration with the VNIITE department of ergonomics, and was produced by the Onezhsk Tractor Plant over the course of nearly four decades from 1966 to 2003.

The 'Next Generation Taxi', launched by the VNIITE
in 1964

In 1962, the USSR Council of Ministers issued a resolution calling for the creation of a specialised taxicab. The result was the Soviet 'Next-Generation Taxi', launched by the VNIITE in 1964. In the course of their preliminary research, specialists in the field of motor vehicles established that the work conditions for a taxicab were not the same as those of a general-purpose automobile. Industrial designers Yuri Dolmatovsky, Alexander Olshanetsky and their colleagues worked on developing the vehicle's design. Resembling a small bus, the NGT was reliable, manoeuvrable, compact yet capacious, with an even floor along the body, an enclosed trunk, sliding doors and a separate cab for the driver. (A disconnectable cooling fan was also a feature, the ducts and wiring installed along the left side of the vehicle.) While only two prototypes of the taxi were made, one of them was in operation on the streets of Moscow for a month. In terms of the imaginative solution and basic functional and performance indicators, this project was no less successful than those proposed a decade later by the famous Italian designer, Giorgetto Giugiaro. Nevertheless, for a number of reasons the NGT was never mass-produced.

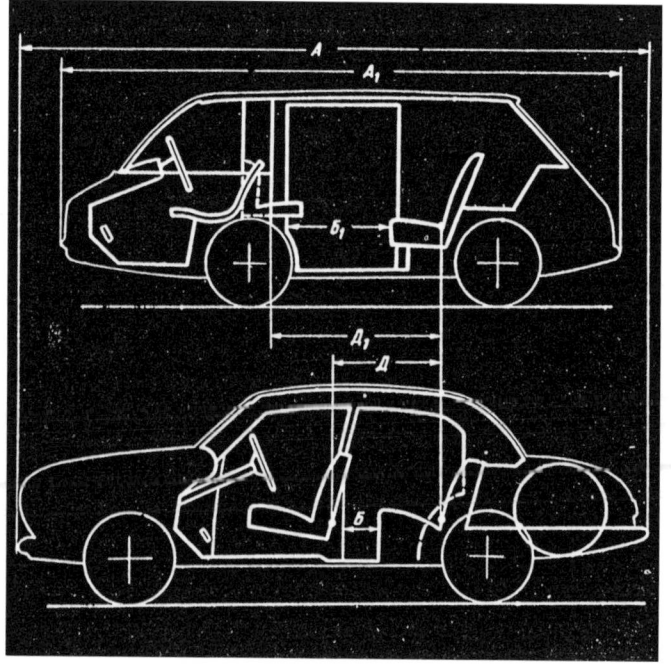

Design plans for the 'Next Generation Taxi'

The taxi had sliding doors and a separate cab for
the driver (see over page)

The truck's designer, Valentin Kobylinsky

The BelAZ-540 quarry truck
(model shown, over page)

In 1965, the VNIITE industrial designer Valentin Kobylinsky proposed a design for the BelAZ-540 quarry haul truck that allowed the load capacity of the vehicle to reach 27 tons while ensuring that it was easy to operate. During use, regular windscreens quickly became covered with dust, so Kobylinsky designed the truck's cab with the glass tilting away from the driver. For extra protection the designer also extended the edge of the truck's body, effectively covering the cab with a roof. Kobylinsky was the first designer to introduce hydropneumatic suspension for the wheels, alongside a general hydraulic system for the power steering and the body hoist. The first BelAZ vehicles had four headlights, but in the newer model the designer proposed placing a unit with six lights along the truck's front edge. The vehicle was designed in one year and was awarded a gold medal at the international exhibition in Leipzig. Once production began the truck was exported and the characteristic configuration, with the driver's cab shifted to the left, is still used in equipment issued by the BelAZ plant. In recognition of the high quality of his work, Kobylinsky was awarded the Order of the Badge of Honour. The truck itself became a symbol of large-scale housing construction in the USSR.

The Sign of Quality mark, designed by Valeri Akopov
in 1967

The role of the VNIITE included examining the consumer-related properties of manufactured goods, evaluating the aesthetics of new products and certifying them in accordance with the highest quality ranking. In 1967, in order to boost the quality and efficiency of manufacturing, the government-approved Sign of Quality campaign was introduced. Its logo was developed by graphic designer Valeri Akopov and was issued for a time period of two to three years following the results of government certification. It was applied to products, or their packaging, and also placed on shipping documentation, tags and labels. An item could only receive a Sign of Quality mark if it passed a VNIITE expert review. Special attention was given to those goods that were the most complex in technical terms, such as tape recorders, refrigerators and motorcycles. Receiving approval for the development of technical specifications by the VNIITE allowed adjustments to be introduced at the earliest stages – only the highest quality designs would be allowed to be mass-produced. This way, the design had an impact on product development during every step, from formulating the objective to producing a sample for mass production.

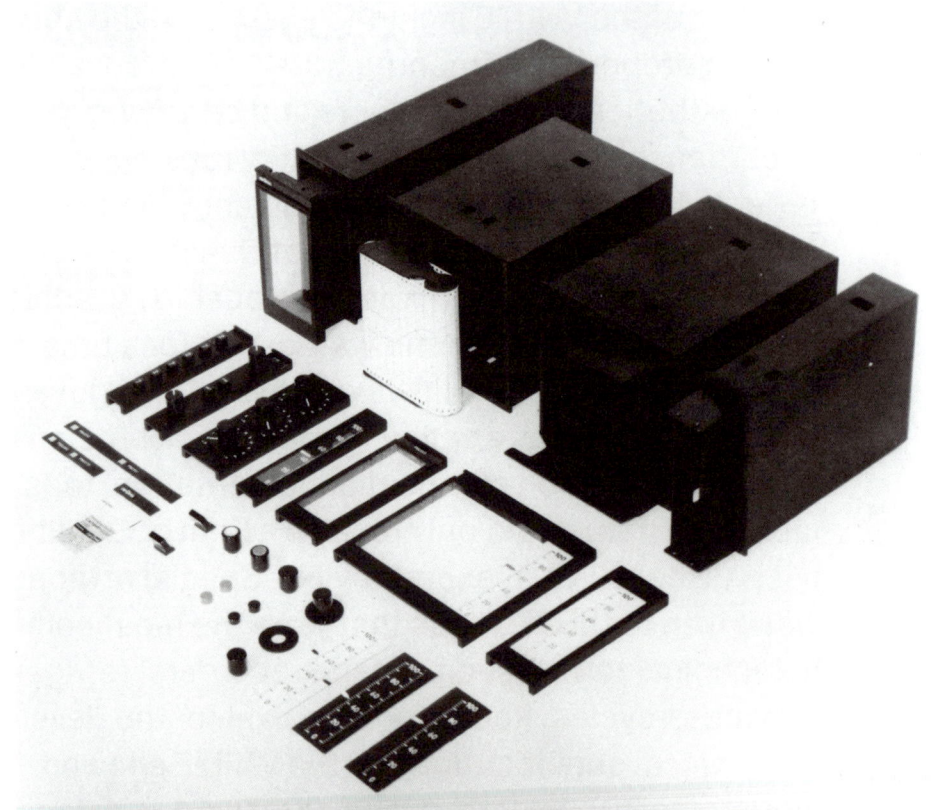

Control panels for measuring instruments,
designed as part of the VNIITE's Elektromera
programme (and over)

A design programme represents a method of designing goods and managing project work. For the VNIITE, these programmes were focused on the formation of large multidisciplinary products that had a single function. One of the first such projects was the Elektromera programme, which was developed by Dmitry Azrikan, Alexander Grashin, Lev Kuzmichyov, Ramyz Guseinov and other VNIITE employees from 1974 to 1979. The Soviet Union produced more than 1,200 types of electrical measuring instruments and the VNIITE industrial designers created a universal set of components for control panels and body parts. A system was developed for consistent visual information, including instrument lettering and signs that are still in use today; standardised solutions for the workplace environment and work gear; and packaging design. Using the Elektromera allowed the quantity of measuring devices used at factories to be reduced by a factor of four and the labour involved in producing structures was also reduced dramatically. Concomitantly the ergonomic and aesthetic properties of various equipment was improved and a unified organisational style was created. In 1982, the programme was awarded a prize by the USSR Council of Ministers and the standardised rules for configuring control panels remain relevant today.

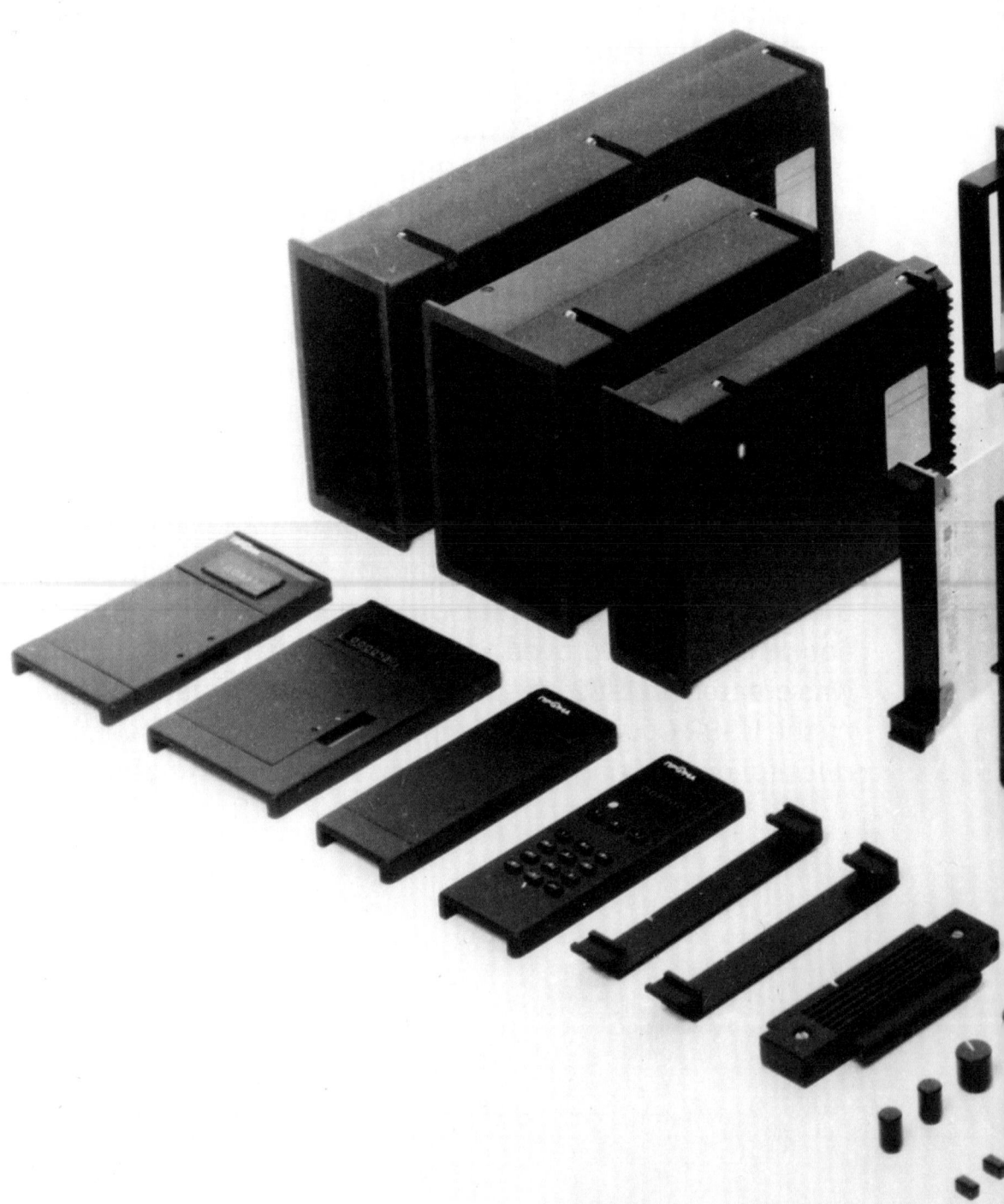

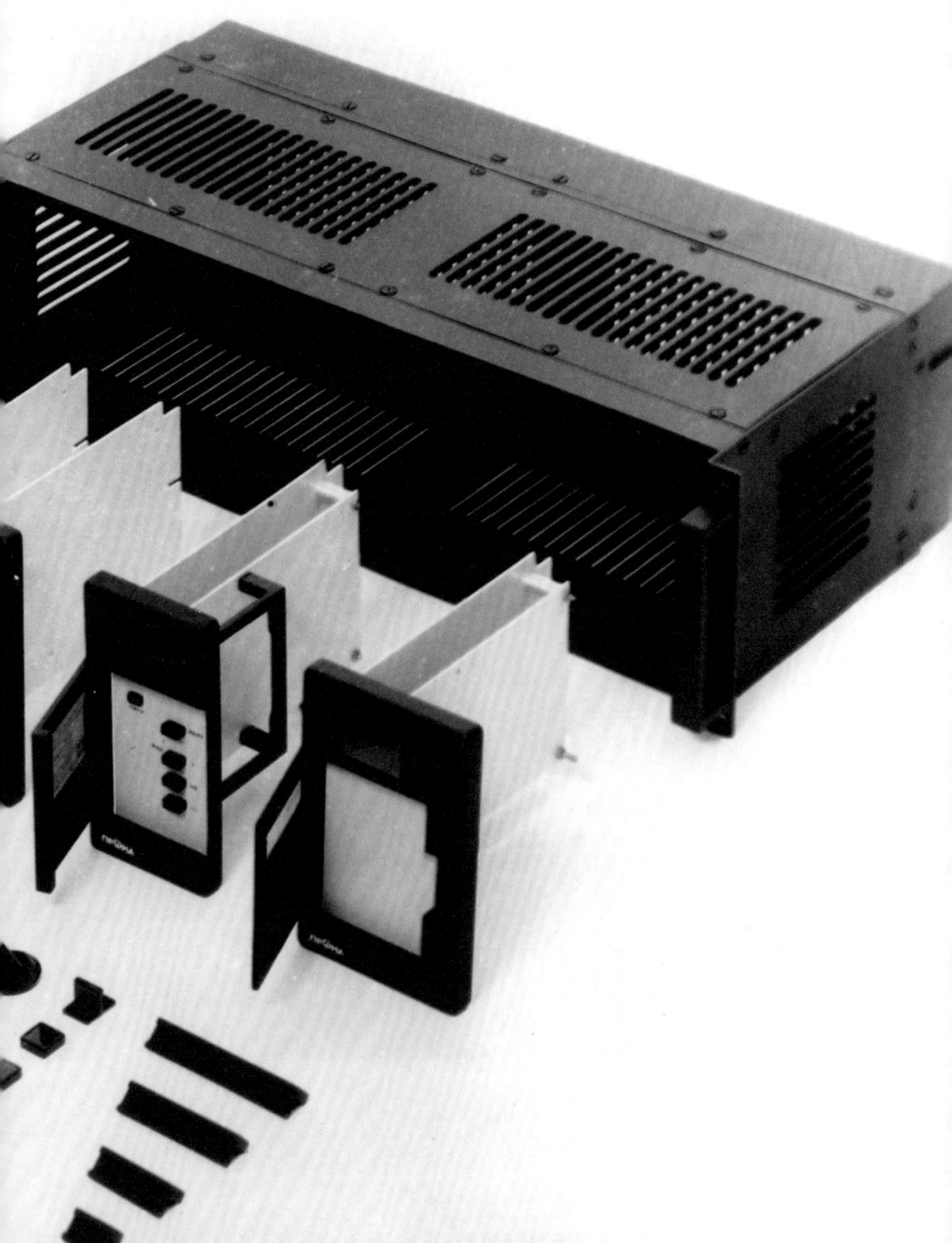

The VNIITE's Alexander Grashin (standing, at centre)
and prototype for a hydraulic copy lathe, designed
by the institute for the Utita company (and over)

In 1975, a foreign company ordered a VNIITE project for the first time: a semi-automatic, hydraulic copy lathe with a programmable controller. The machine tool was manufactured by the Italian company Utita and had a number of serious deficiencies: namely, it was inconvenient to use, adjust and clean and its dimensions were large. The designers were given the task of improving the technical, aesthetic and operational quality of the machine tool, as well as its look (the form, colour scheme solution and graphics), making the model more competitive by doing so. A considerable part of the early project studies involved analysing the production process. As a result the designers identified areas where the machine tool operator was subject to a high level of physical stress (on the muscles and eyes) while working – and these needed to be adjusted. A comparative analysis of the machine tool prototype and the new version was conducted on a series of full-sized models. Specialists in the VNIITE machine tool department, led by the industrial designer Alexander Grashin, managed to remedy the defects, optimise the machine's ergonomic characteristics and increase its productivity. Utita gave the work a positive assessment and the Italian media ran reports on a successful collaboration with Soviet designers.

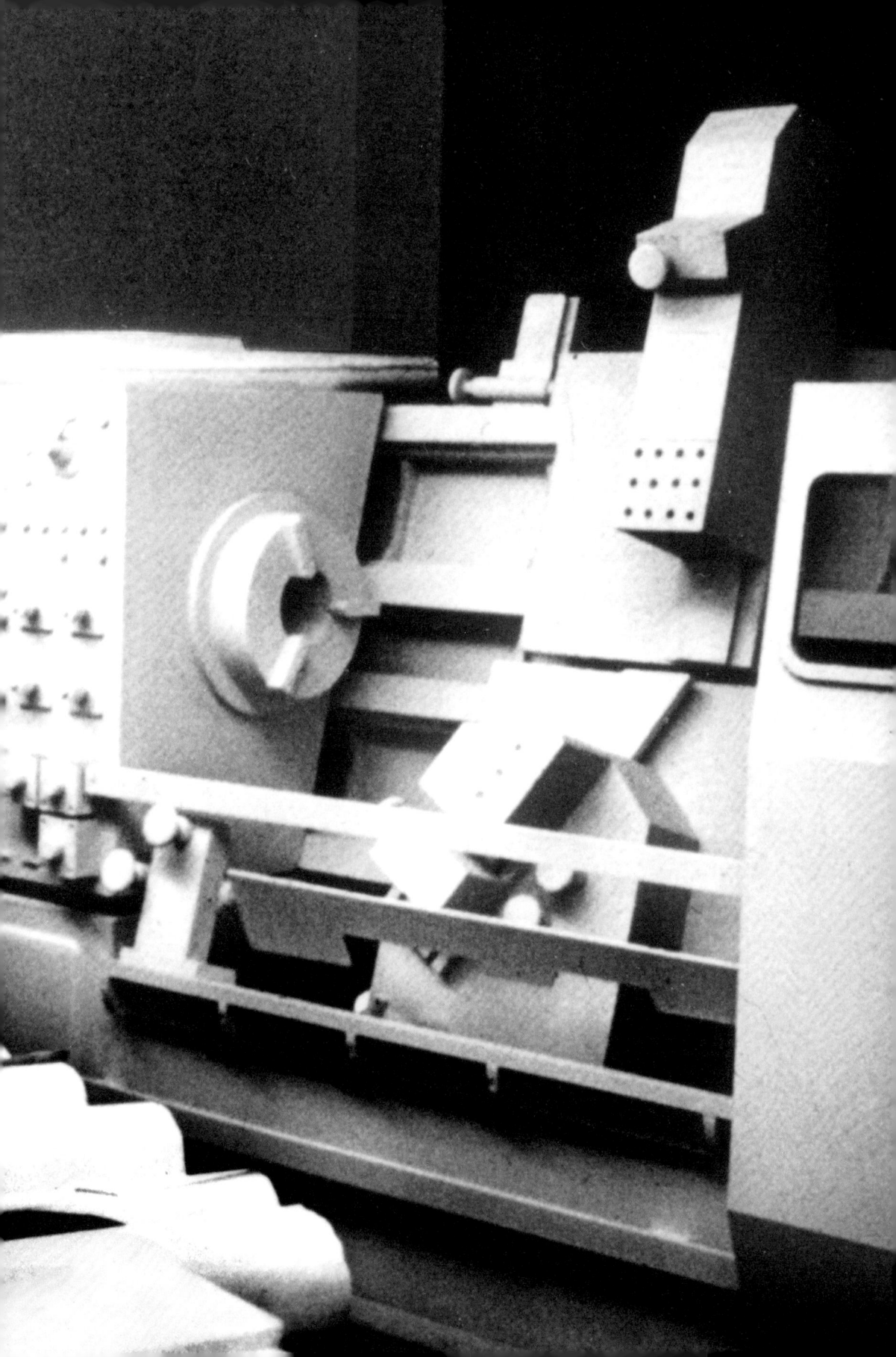

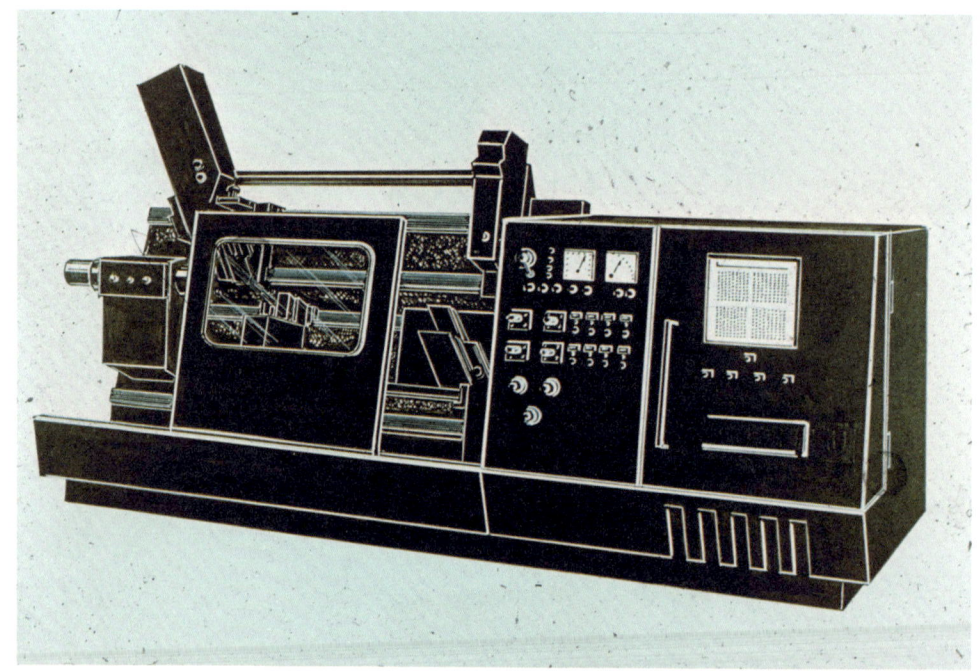

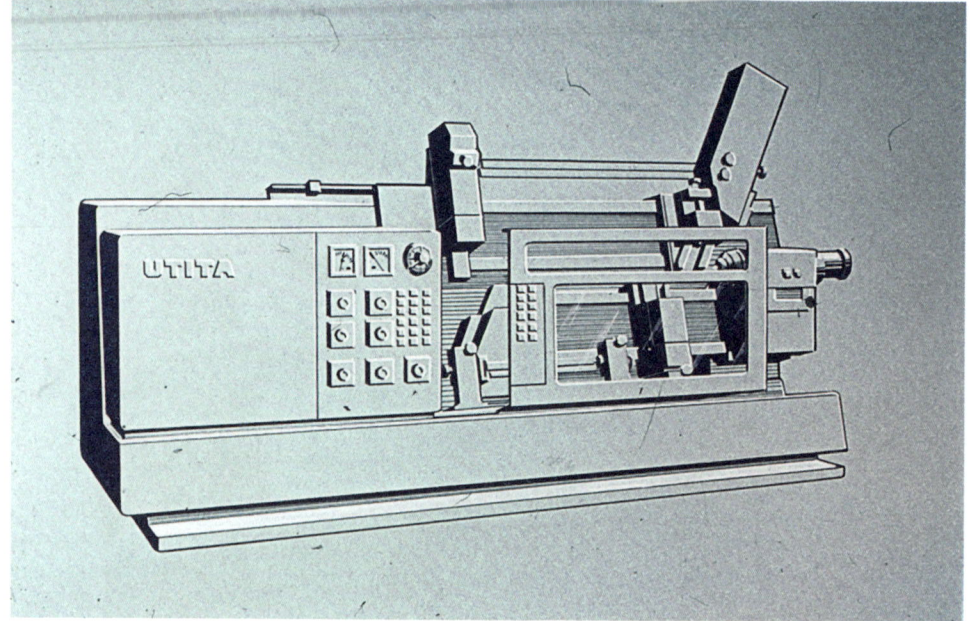

Various design drawings for the VNIITE/Utita lathe

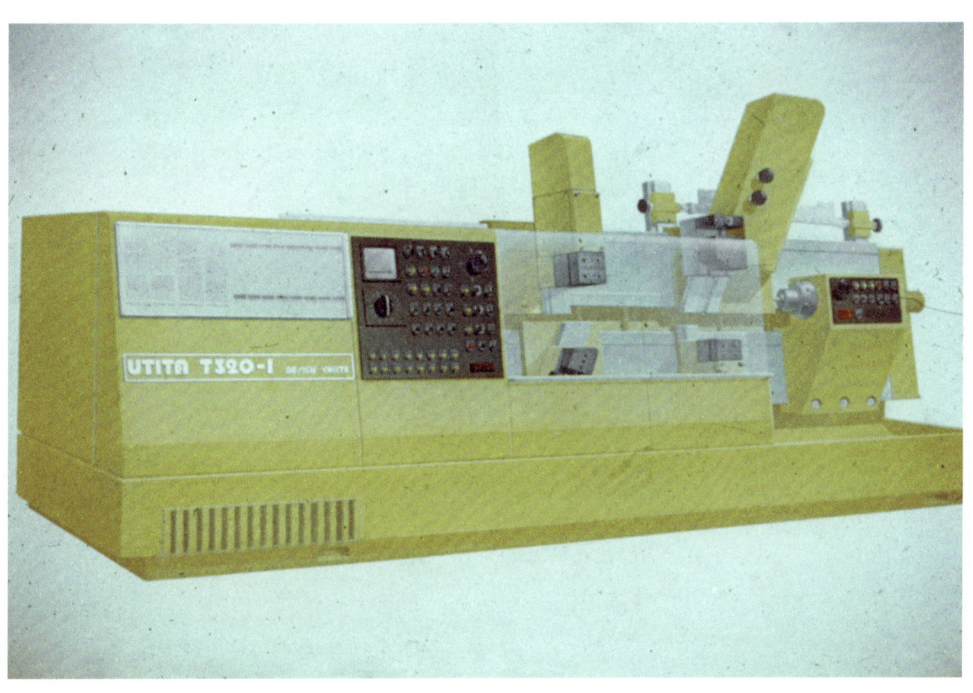

Models of the ZIL-Sides fire engine

In the early 1970s, the VNIITE was assigned the task of creating a fire engine using new design solutions and mass-produced domestic components. It was necessary to provide for comfort but also ensure that firefighters could board and disembark the cab quickly. The first prototype of a fire engine using the carriage configuration that was popular in those years was made in 1973 in the pilot construction design office for fire department vehicles. In 1975, joint work was started between Vneshtekhnika (USSR) and the Sides company (France) to develop the new vehicle – the scale models shown here pertain to the ZIL-Sides BMA-30 fire engine conceived by industrial designers Vladimir Aryamov, Lev Kuzmichyov, Alexander Olshanetsky and Tatiana Shepelyova (Sides adhered to the project design proposed by the VNIITE). Two pilot vehicles were assembled in Saint-Nazaire, France and these prototypes were featured in numerous exhibitions before being used at the various stadiums during the 1980 Olympic Games in Moscow. Nonetheless, the vehicle proved to be too expensive and did not fit well into existing archetypes, so was never mass produced.

Attendees at the IX ICSID Congress which took
place in the Rossiya hotel in Moscow, 1975

In 1965 the VNIITE became a member of the International Council of Societies of Industrial Design (ICSID). The images on the following pages show the IX ICSID Congress devoted to the topic 'Designing for people and society' that was held in the Rossiya hotel in Moscow in 1975. This was the first ICSID Congress to be held in a Socialist country; around 1,500 specialists attended, including 750 foreign specialists from 32 countries. The Congress was an important event for Soviet design and it was necessary to select the best developments in design from around the USSR and find the most appropriate ways to present them. Different areas were created to hold business meetings and discussions and were equipped with cardboard furniture made according to drawings produced by the architect Alexander Ermolaev. In 1977, the VNIITE's founder and Director Yuri Soloviev was elected President of the Council (the only time ICSID was led by a Soviet designer). Over the course of his long career, Soloviev was elected Vice-President three times.

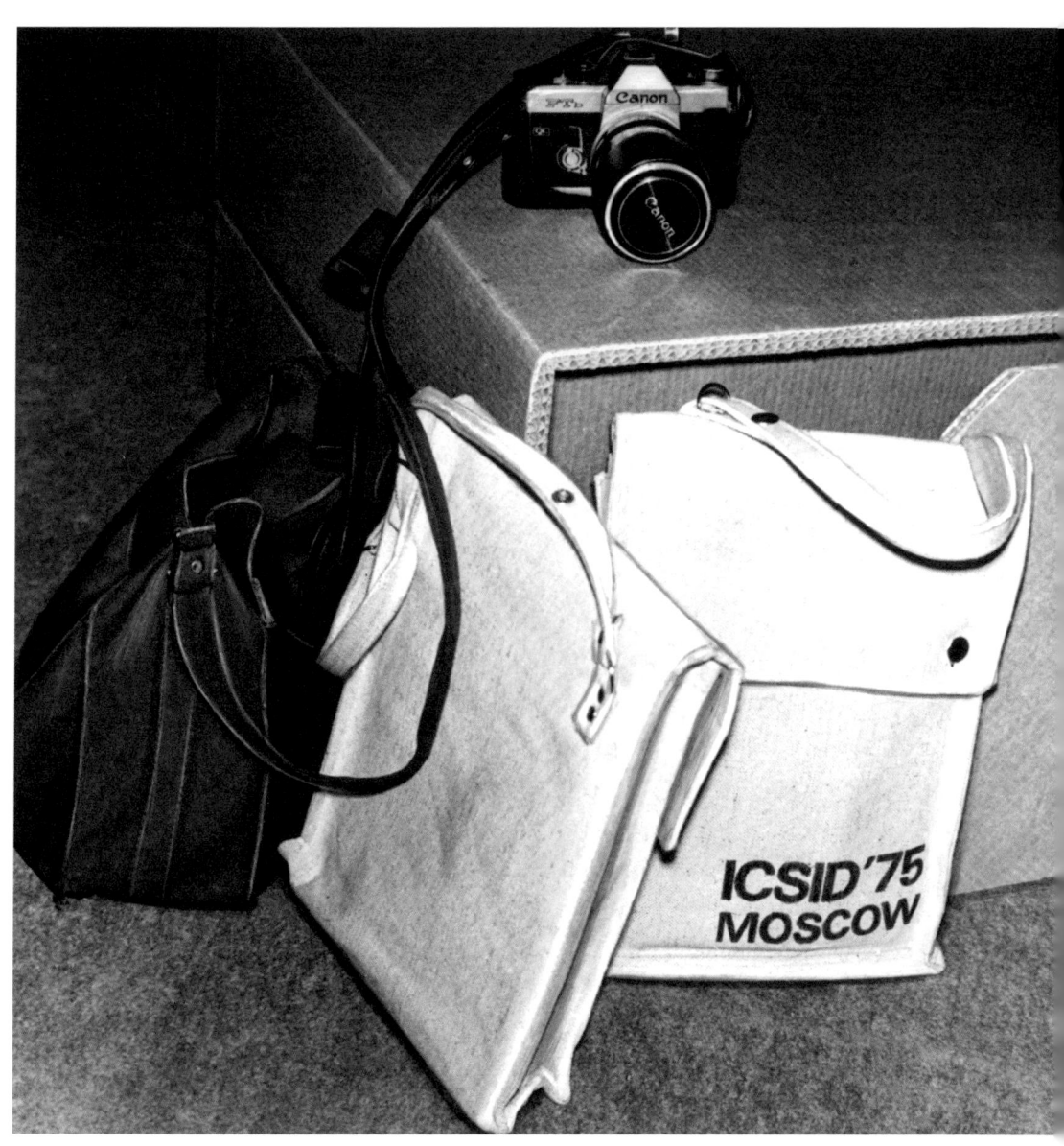

The IX ICSID Congress in Moscow also
featured cardboard furniture designed by
Alexander Ermolaev

The high-speed tramway car designed by
the VNIITE in 1976

The design for a high-speed tramway car shown here was created in 1976 by the Urals branch of the VNIITE together with the Ust-Katav Rail Car Plant. A team composed of the industrial designers Stanislav Zaritsky, Valery Berdyugin, Alexey Fedotov and colleagues worked on the design. The four-axle, all-metal car could gain speeds of up to 80 kmph, carry up to 200 passengers across one, two or three cars, and was intended for urban, suburban, and intercity routes. Compared to previous models, the method of entering and exiting the cars and the space for passengers was optimised by dividing the compartment into two parts – for those traveling long or short distances. The passengers' view was increased by enlarging the width of the window openings as well as lowering the bottom edge of the windows themselves. The electrical equipment and contact panels were transferred to the tramway car's roof, allowing for a reduction in the length of electric power utility lines, as well as improving conditions for installing and maintaining the cars. The new heating and ventilation systems enabled the car to operate up to 1,200 metres above sea level and at ambient temperatures ranging from -40°c to 40°c.

The VNIITE's exhibitions on Soviet design took place all over the world, including India (overpage)

Tractor model (and various levers) on display at a VNIITE show

The VNIITE was active on an international level and organised exhibitions of Soviet design abroad that displayed goods produced by industry, scale models and prototypes. In 1976, in Stuttgart the 'Soviet Design' exhibition was organised by the VNIITE and became the first large-scale demonstration of design from the USSR in West Germany. The exhibition included three main categories: 'Formation of design in the USSR', which demonstrated goods created by Soviet industrial designers and noted the prospects for developments in design; 'The methodologies in Soviet design' and 'Training personnel'. One of the distinguishing aspects of the exhibition was that visitors could not only see finished goods, but could observe the different stages in the process of industrial design and become more familiar with the concepts and work methods of designers (this was of particular interest to foreign professionals). International exhibitions held immense importance for the development of design in the USSR (attendance at Stuttgart numbered more than 5,000 people). They provided the chance to establish business contacts with foreign partners and professional societies, and informed the world of the latest developments at the VNIITE.

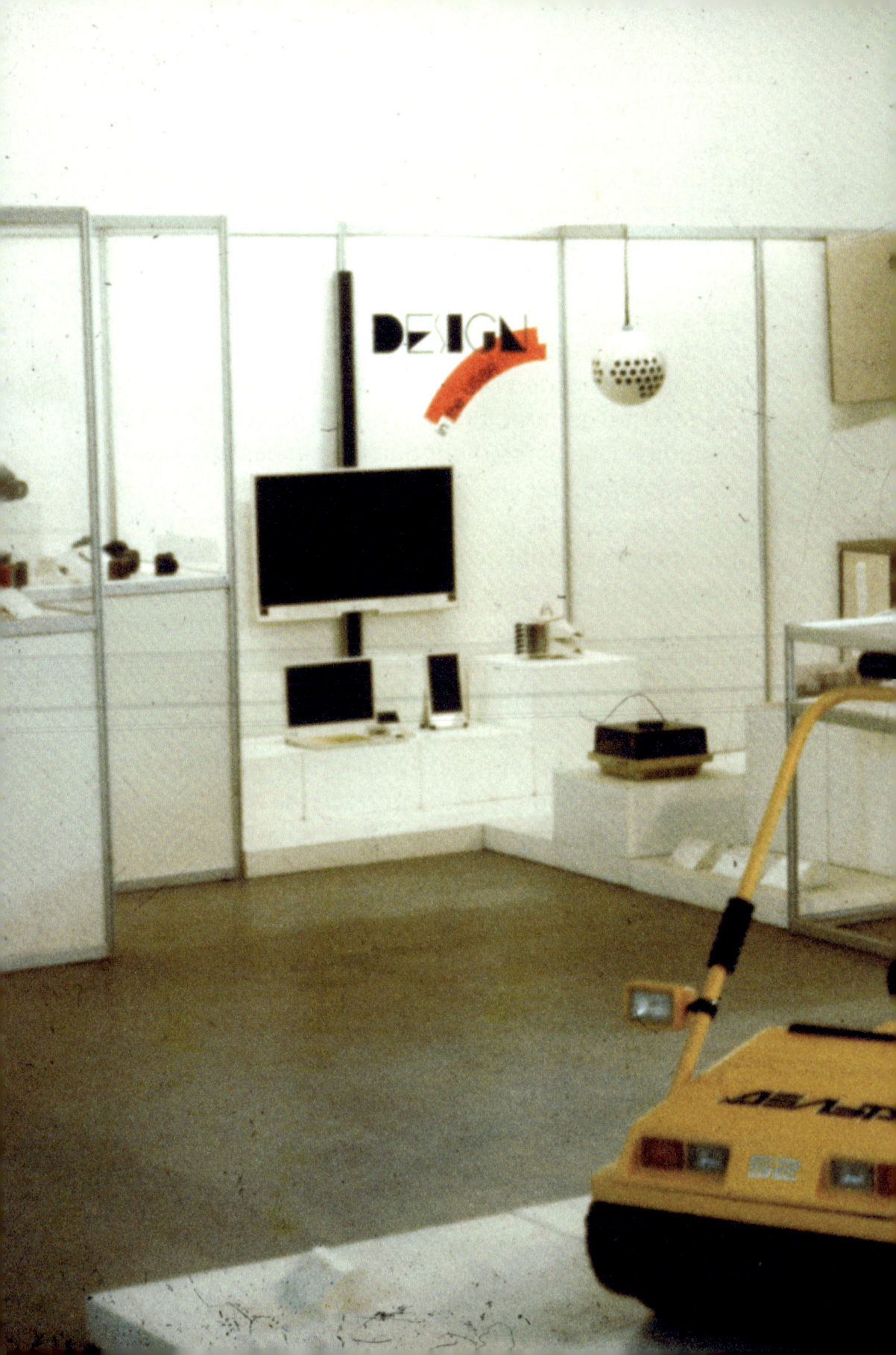

ТЕХНИЧЕСКАЯ ЭСТЕТИКА 1966 12

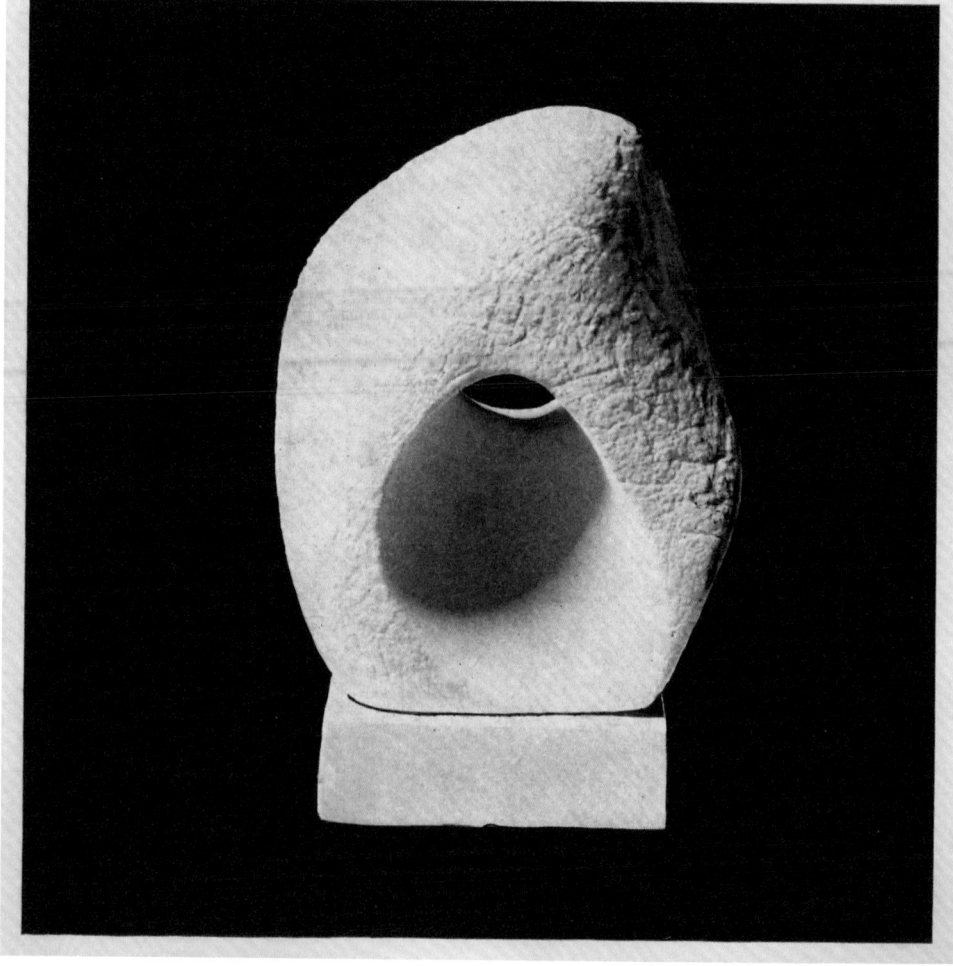

Issue 12 (1966) of the VNIITE's monthly journal,
Technical Aesthetics

The VNIITE journal, Technical Aesthetics, was published monthly from 1964 to 1992 and then irregularly until its final edition in 1995. The magazine was very popular among a professional audience and published articles on the theory and history of design while addressing contemporary issues concerned with industrial design and its methodologies. It also covered ergonomics, quality control, standardisation, and how design activity was organised. Scientific research carried out by VNIITE employees was also published in various series called 'Ergonomics', 'Materials and finishing industrial goods' and 'Methodological materials'. In addition, the journal published analytical reviews highlighting the achievements of Soviet and international industrial design, as well as annotated bibliographic indexes on design and ergonomics. A series of books published under the banner of 'The Designer's Library' became an important supplement to the journal and these included major works on the history, theory and methodology of design. VNIITE publications played a significant role in helping students in specialised colleges and industrial designers raise their qualifications. In 1997, the publications were awarded the Russian Federation State Prize for literature and art.

ТЕХНИЧЕСКАЯ ЭСТЕТИКА

1966 **10**

Technical Aesthetics covers from 1967
and 1968

техническая эстетика 1967 **4**

техническая эстетика 1967 7

техническая
эстетика
1969
1

 Technical Aesthetics covers from 1969

техническая
эстетика 1970 8

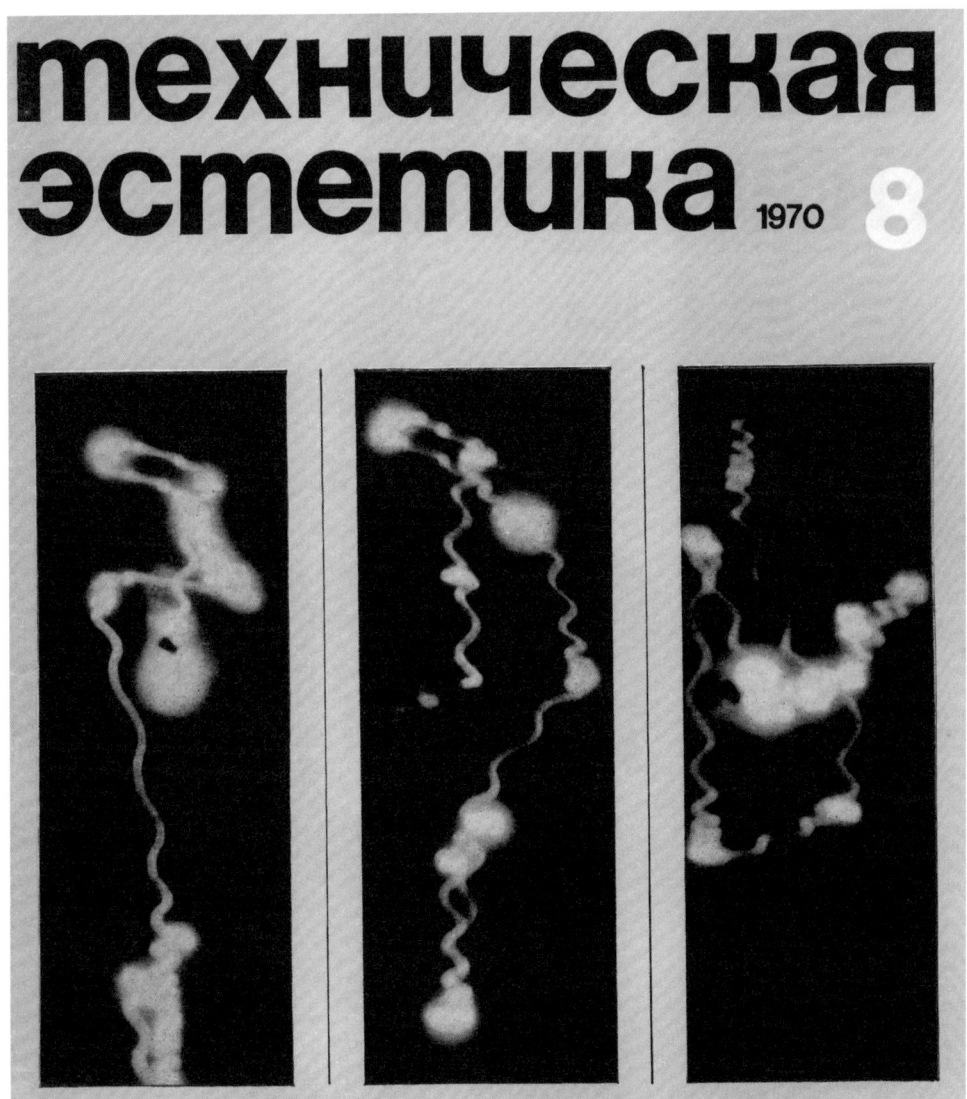

техническая
эстетика
1971 5

техническая эстетика 1972 **2**

техническая эстетика 1972 **7**

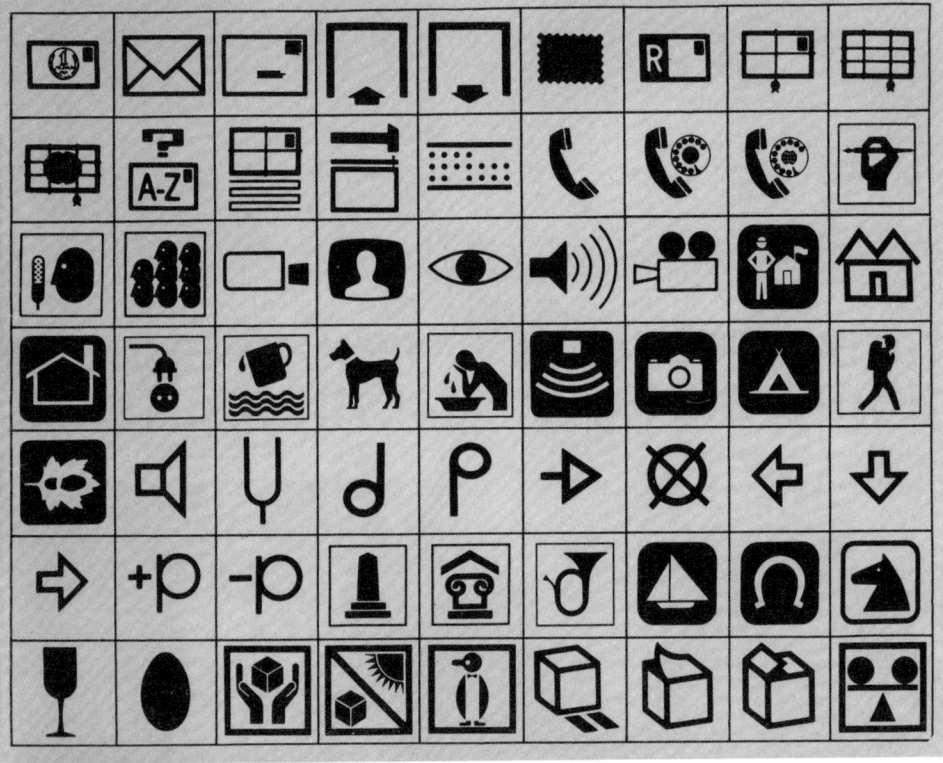

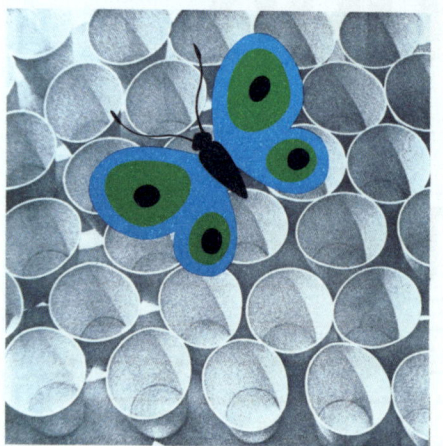

техническая эстетика 1974 **3**

техническая эстетика 1974 **8**

техническая эстетика 1974 **7**

СССР – ГДР

РОЛЬ ХУДОЖЕСТВЕННОГО
КОНСТРУИРОВАНИЯ В
РЕШЕНИИ КОМПЛЕКСНЫХ
СОЦИАЛЬНО ЗНАЧИМЫХ
ЗАДАЧ ПРОЕКТИРОВАНИЯ
ПРЕДМЕТНОЙ СРЕДЫ

техническая эстетика 1974 **9**

техническая эстетика 7

1975

техническая эстетика

1975 **3**

техническая эстетика 1975 11

техническая эстетика
1/1981

ISSN 0196-5369

техническая эстетика

3/1981

ICSD
INTERDESIGN 80
TBILISI

Technical Aesthetics covers from 1978, 1982,
1985 and 1989

техническая эстетика

3/1985

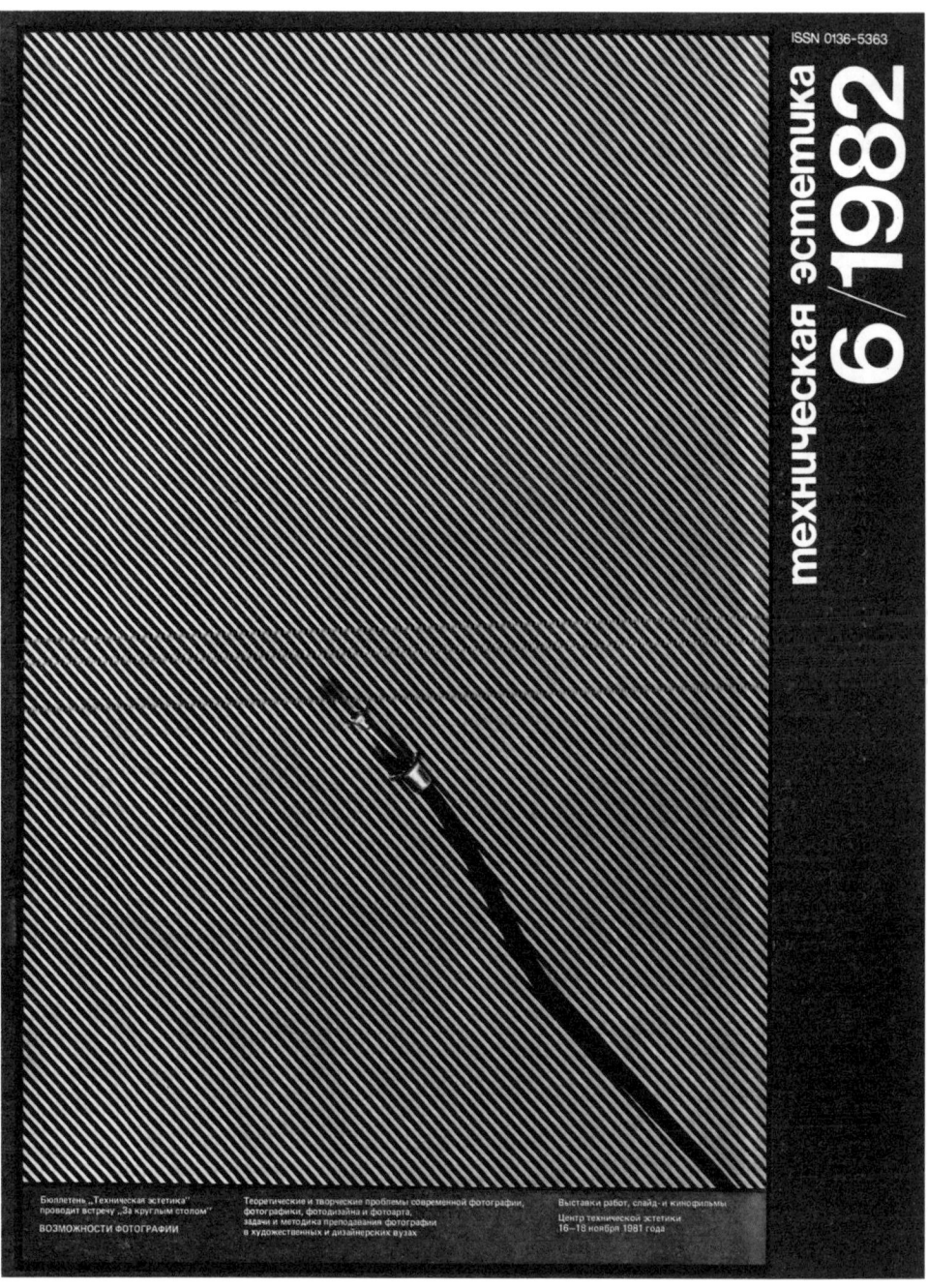

ISSN 0136-4898

техническая эстетика
10/1982

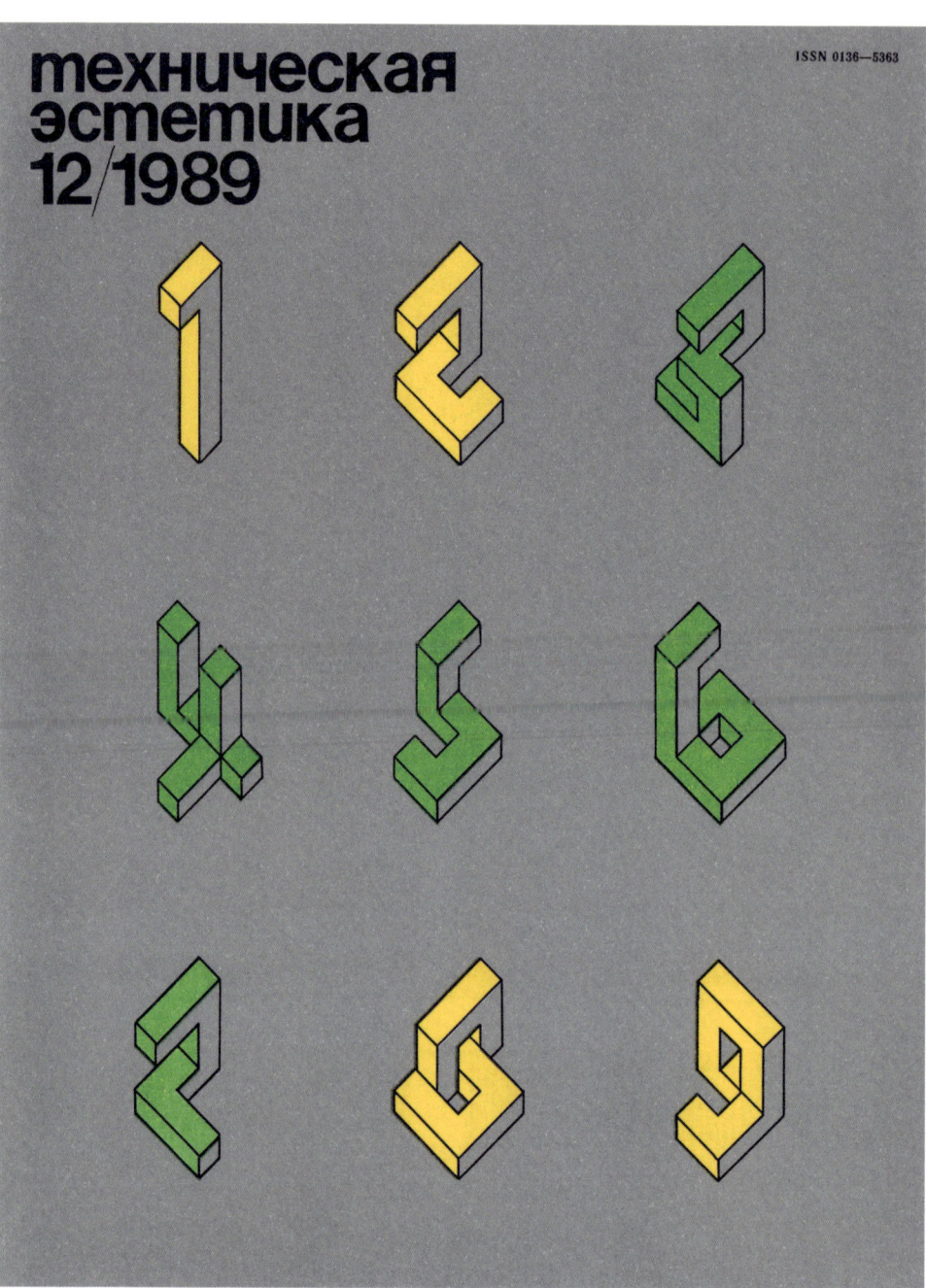

ISSN 0136—5363

техническая
эстетика
12/1989

Technical Aesthetics covers from 1989 and 1990

ISSN 0136-5363

техническая эстетика   6/1990

ISSN 0136-5363

техническая эстетика

2/1990

ISSN 0136—5363

№7(331)1991

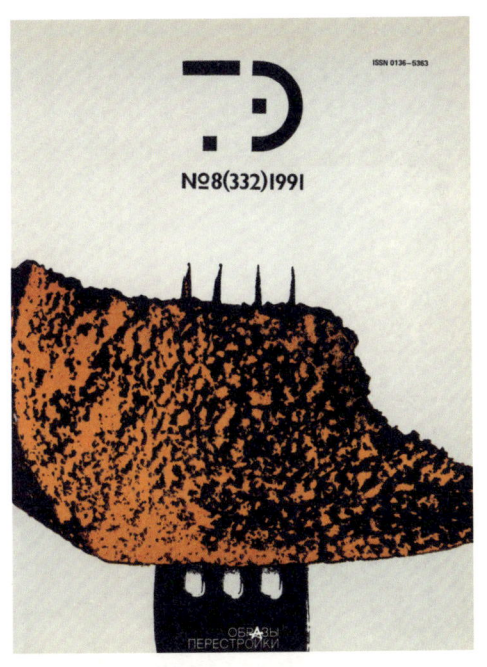

№8(332)1991

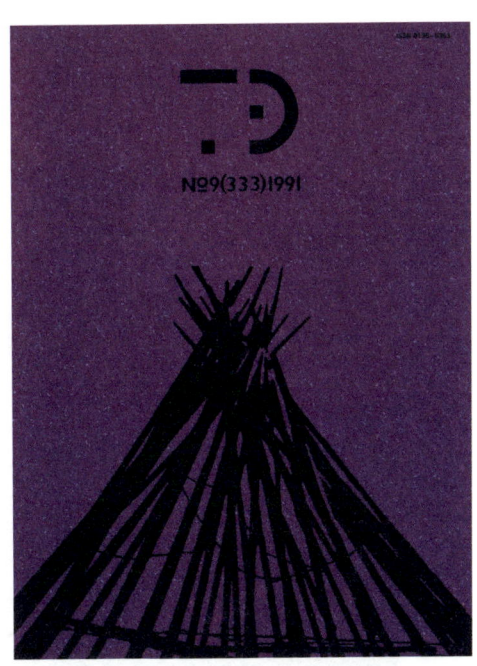

№9(333)1991

№10(334)1991

№11(335)1991

The windows of the Centre for Industrial
Design Aesthetics displayed new VNIITE projects
and featured screens showing demonstrations

In 1977, as part of the popularisation of design methods used in industrial production, the Centre for Industrial Design Aesthetics (CIDA) was opened in Tverskaya Street in Moscow. Its work was focused on two areas: the promotion of artistic design methods and the design practitioners themselves. Seminars, lectures, and awareness-raising days were held for the employees of various manufacturing facilities. The Centre's display windows showcased operational samples of goods and large-scale equipment models, allowing residents and visitors to the nation's capital to become familiar with the best of the domestic design work being produced. Having constructed a row of screens, VNIITE specialists created films to accompany the exhibitions. For example, if kitchen equipment was displayed, an actor on the screen would talk about the product and demonstrate how convenient it was to use. Once a quarter, a new exhibition would be launched and receive coverage in major publications in the capital. The CIDA's work facilitated the popularisation of artistic design methods, established a professional environment, increased the educational level of various specialists, and forged and strengthened relations inside the design community.

VNIITE designers in front of images of
various consumer goods produced by the
Kultbitmash programme

In 1977, the State Committee for Standards and Product Quality Management (Gosstandart) adopted a resolution called 'On using systemwide industrial design work in integrated standardisation programmes'. From this point onwards, having an aesthetic solution for goods was laid down in the foundations of regulations governing mass production. The first design programme to increase the quality of goods in high demand was Kultbitmash, developed by the VNIITE and its branches for the Ministry of Machine Tool Building. The images on the following pages show the range of consumer goods to which these new standards were applied: sewing machines, irons, telephones, bicycles, vacuum cleaners, fans, children's strollers, tape recorders, watches, electric shavers, and more. The development was unified by a universal approach to forming name classifications for goods (a common method and principle in industrial design), and a means of applying colour graphics across the board. The design programme was conceived in three steps – creating the concept; developing name classifications and product mixes; and directly implementing the solutions into the production process – and remained in use until 1985.

Products being evaluated at a Kultbitmash design
programme meeting

The Kultbitmash programme focused on improving the design of goods that were in high demand, such as vacuum cleaners and (previous page) sewing machines

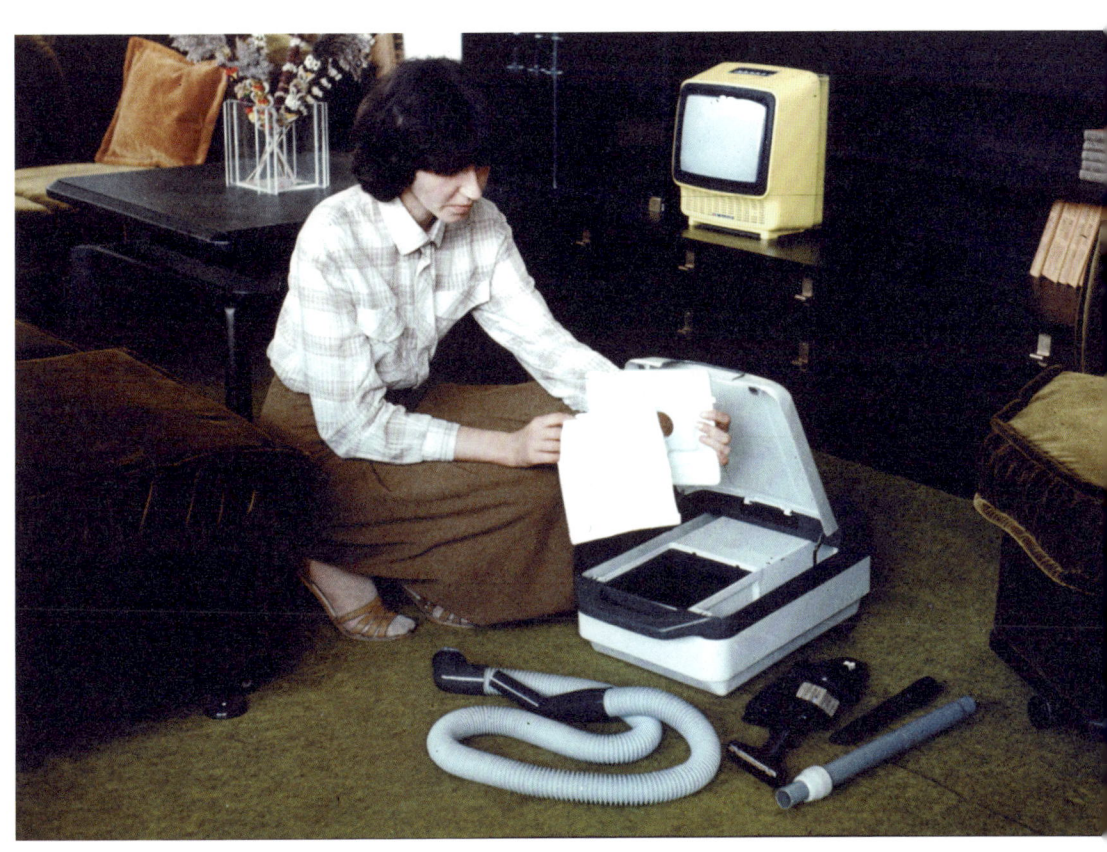

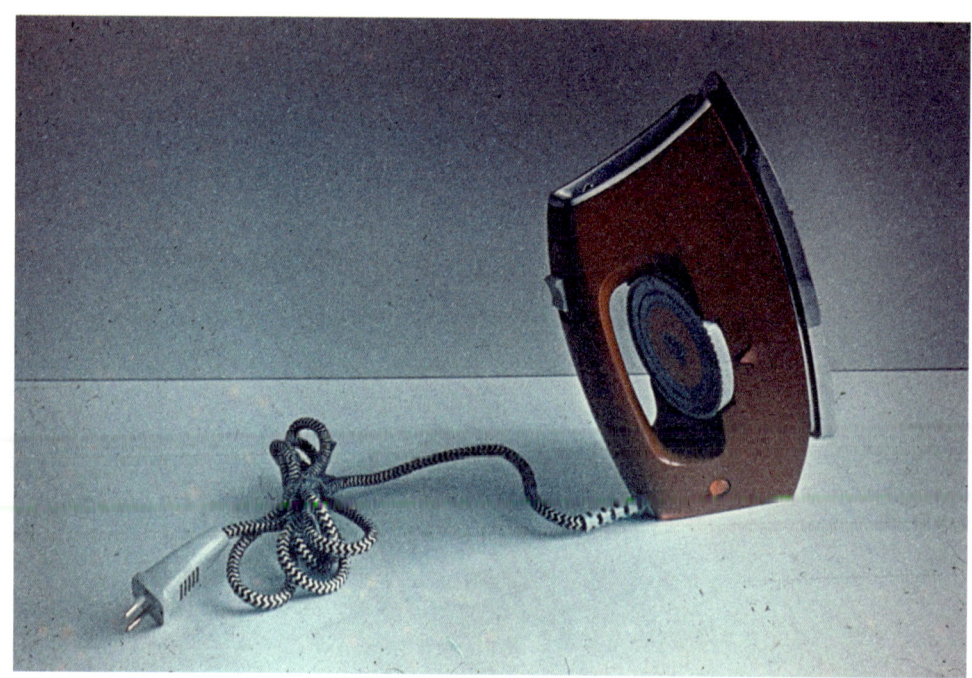

Iron, hair dryer and (overpage) telephones produced
by the VNIITE's Kultbitmash design programme

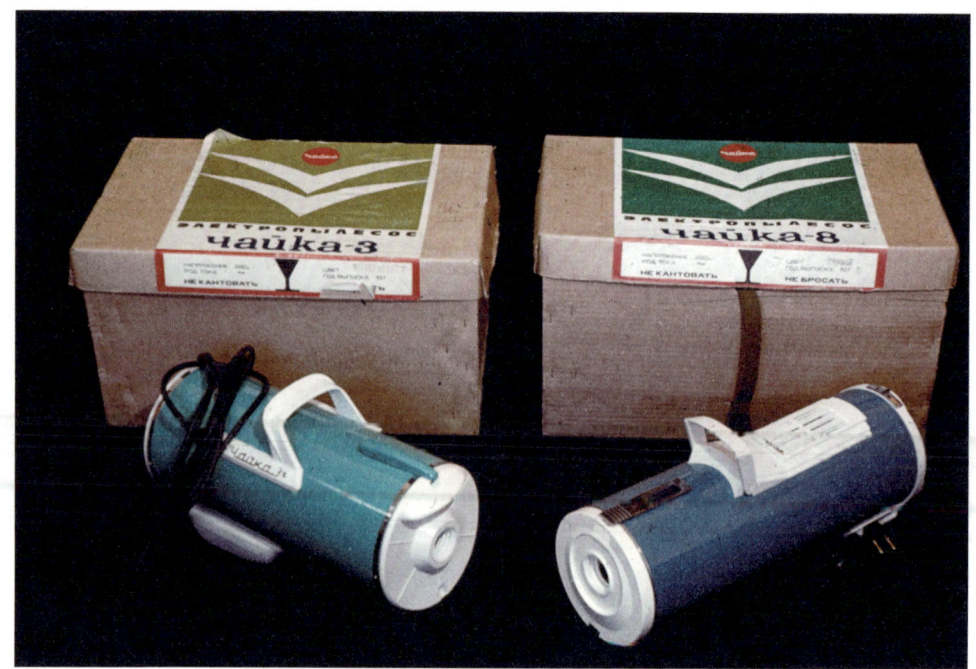

Products from the Kultbitmash design programme
(sewing machine shown overpage)

The Interdesign-77 seminar held at the VNIITE
Kharkov branch addressed the topic of
designing for elderly and disabled people

Scientific research was carried out not only in Moscow, but in the VNIITE's regional subdivisions. In 1977 a project seminar called Interdesign-77 was held at the VNIITE Kharkov branch, organised together with the International Council of Societies of Industrial Design (ICSID). Usually, these kinds of seminars took place over the course of two weeks and their purpose was to address solutions for social problems. The Kharkov Interdesign seminar was attended by 27 specialists from the USSR and 11 countries: the UK, Hungary, East Germany, West Germany, Ireland, Italy, New Zealand, the USA, Czechoslovakia, Sweden, and Japan. The organisational committee's Chairman was VNIITE Director Yuri Soloviev, while the English designer John Reid was the seminar's Coordinator. The topic the professional community was invited to focus on was designing for the disabled and the elderly. The participants approached the problems by designing household and industrial equipment for people with disabilities – including kitchen equipment, a work chair, a city commuter bus, and a workplace for a seamstress.

Classes at the VNIITE's Department of Theory and History of Design. Selim Khan-Magomedov is shown in the bottom photograph, on the left

The Department of Theory and History of Design at the VNIITE was one of the strongest in the world. Alongside industrial designers, many other professionals worked at the school including art and design historians, philosophers, cultural experts, sociologists, psychologists and architects. From 1978 to 1988 the Department conducted weekly theoretical seminars entitled 'The artistic problems of the object-spatial environment'. The classes were held under the leadership of Selim Khan-Magomedov, a well-known art scholar, historian and researcher of the Russian avant-garde. Within these theoretical seminars, participants reviewed the philosophical, cultural and artistic aspects of the interaction between art and design and discussed different concepts for project work. Over 350 sessions and 30 scientific conferences were held during this time, in addition to almost 700 lectures that were published in specialised print media. Among those who delivered talks were several Soviet design pioneers; the first graduates from schools such as the Higher Artistic Technical Studios and the Higher Artistic Technical Institute; artists from the second wave of the Russian avant-garde of the 1960s and 1970s; designers, architects and specialists from various fields in the humanities.

VNIITE designer Nikita Kaptelin working on plans for
the 1980-E snowmobile

The VNIITE was involved in developing transportation for the winter's impassable roads and, towards the end of the 1970s, its designers were some of the first in the world to suggest producing snowmobiles with enclosed cabs. Reports were produced from tests on domestic and foreign snowmobiles (as were surveys of people who drove these vehicles). The main challenge for designers was to make this mode of transport as comfortable as possible. The 1980-E experimental snowmobile with an enclosed cab was developed by Nikita Kaptelin and Alexander Popov in 1979. The nature of the cab permitted travelling at greater speed, ensured safety and comfort, and solved the problem of soundproofing. Uniquely, fibreglass was proposed instead of steel, which meant corrosion (the result of snow and tree branches damaging the vehicle's exterior) could be avoided. Similar to the Niva car, and recognised as the USSR's first comfortable off-road vehicle, the 1980-E snowmobile project was halted after the design solution was proposed.

VNIITE designer Alexei Trushkin working on a putty model
of the snowmobile (and with Nikita Kaptelin, below)

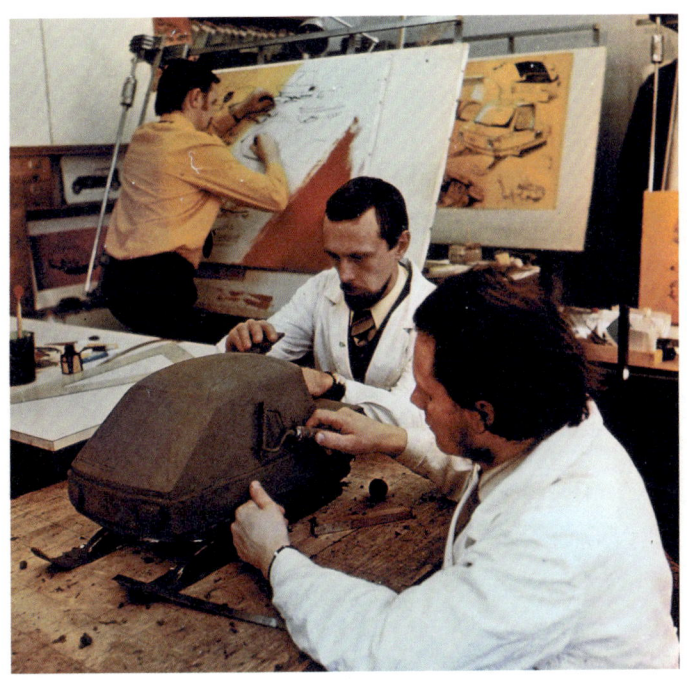

1980-E Snowmobile prototype

VNIITE designers sculpting a putty model

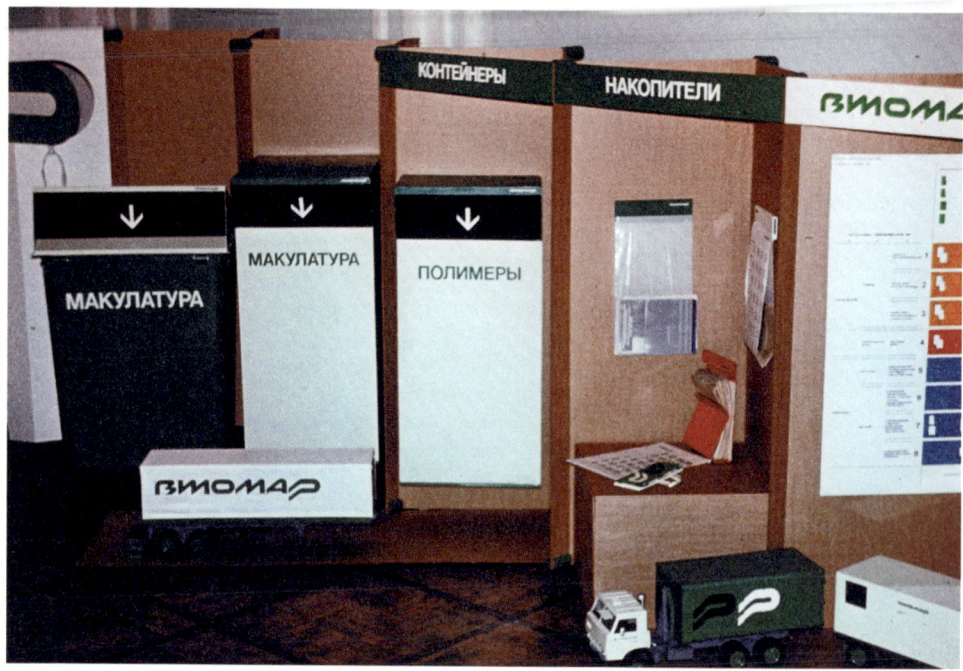

VNIITE team wearing uniforms created for the
proposed VTOMAR waste management system

Identity work and recycling bins designed for
the VTOMAR project

Design programmes that looked at the designing of goods and the management of the particular project became the VNIITE's 'claim to fame' in the early 1980s. One of the most famous programmes of the new, inter-industry type was a project called 'Secondary Material Resources', or VTOMAR, developed from 1979 to 1985 in the VNIITE Leningrad branch under the supervision of designer Andrei Meshchaninov. At the end of the 1970s, the Soviet leadership started to give serious thought to the problem of environmental pollution because a significant amount of household waste was not being returned to the production cycle. The aim of the VTOMAR programme was to improve the system for collecting and recycling recoverable resources. As a result, a new system for collecting recyclables was proposed, based on a method that did not involve incentives. A team of industrial designers developed equipment to collect and transport raw materials, alongside uniforms and various elements for an outreach and awareness-raising campaign. This included a system of colour graphics, designs for thematic exhibitions, advertising and informational materials. An example of a poster highlighting the stages of collecting and processing recyclables is shown on page 133.

Graphics showing how the VTOMAR recycling
system operated

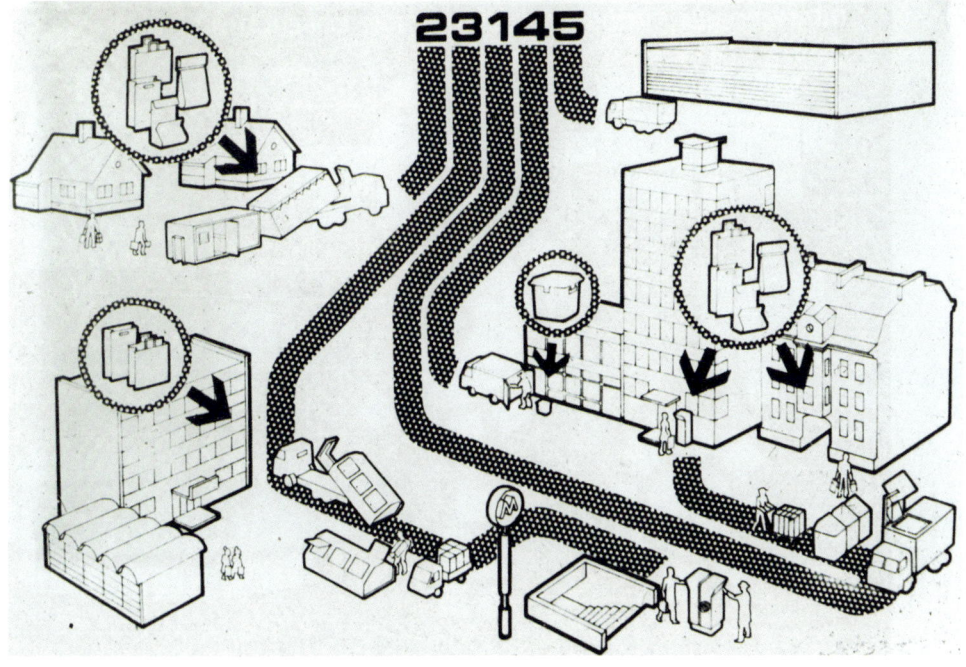

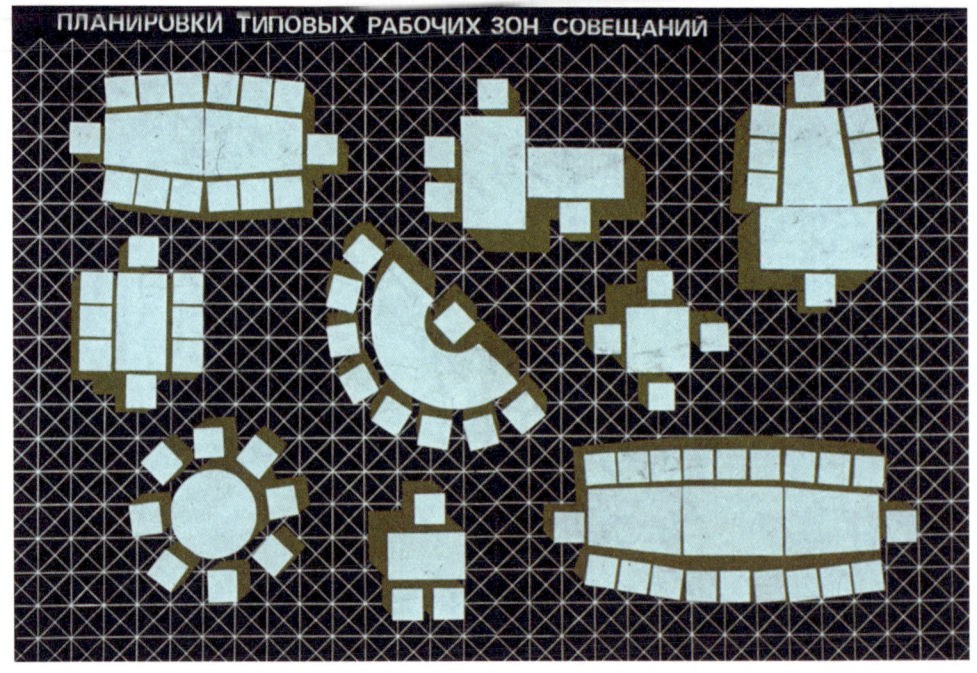

ПЛАНИРОВКИ ТИПОВЫХ РАБОЧИХ ЗОН СОВЕЩАНИЙ

Plans for workplace designs created by the VNIITE
(see overpage)

The design of the workplace was one of the most important areas of focus for the VNIITE. By the beginning of the 1980s, the Soviet Union had accumulated an impressive amount of experience in this area and a corresponding department was created. Later, this area of focus spread across the countries in the Council for Mutual Economic Assistance (COMECON), but its roots were put down in the USSR. The main offices involved with workplace design were the head office in Moscow and the Vilnius branch. VNIITE employees worked on projects for the interior of the Khromatron Plant, the Lenin Komsomol Automobile Plant, the Kama Automobile Plant, and the manufacturing and residential quarters of a Pärnu (Estonia) organisation for cross-industry collaboration. Designers worked together with the production machine operators to choose the colour of the walls and floors in the workshops, the finishing materials and also planned the layout of the machine tools and the exterior environment of the plant, including benches, fountains and pools. The focus eventually shifted to arranging urban space and public areas and, as a rule, designers worked together with architects on these kinds of projects.

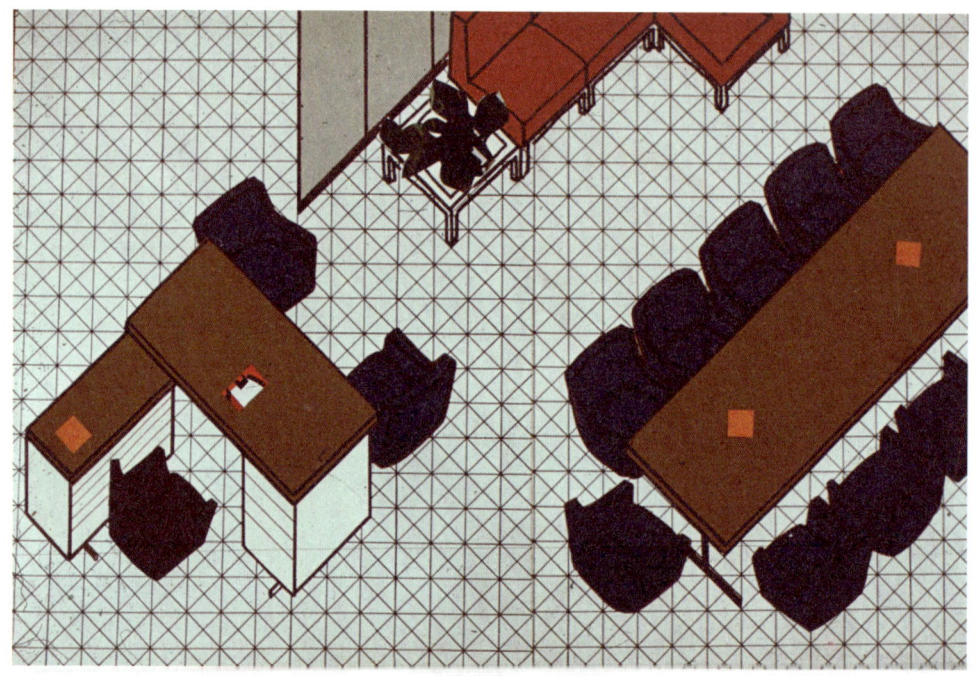

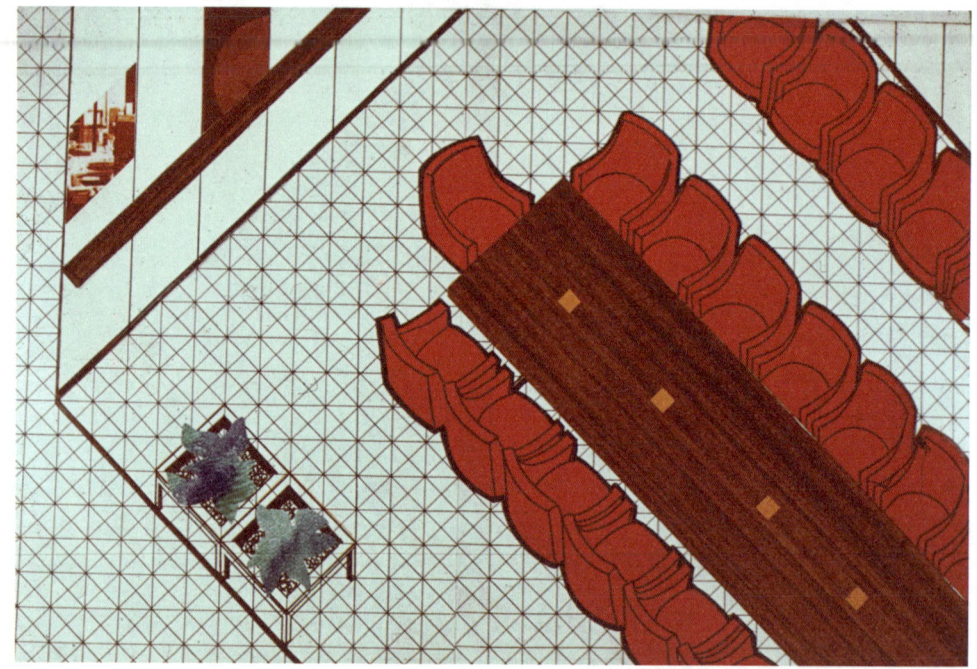

Plans and sketches of various workplace designs
created by the VNIITE (model shown, overpage)

VNIITE designers working on models for
workplace designs

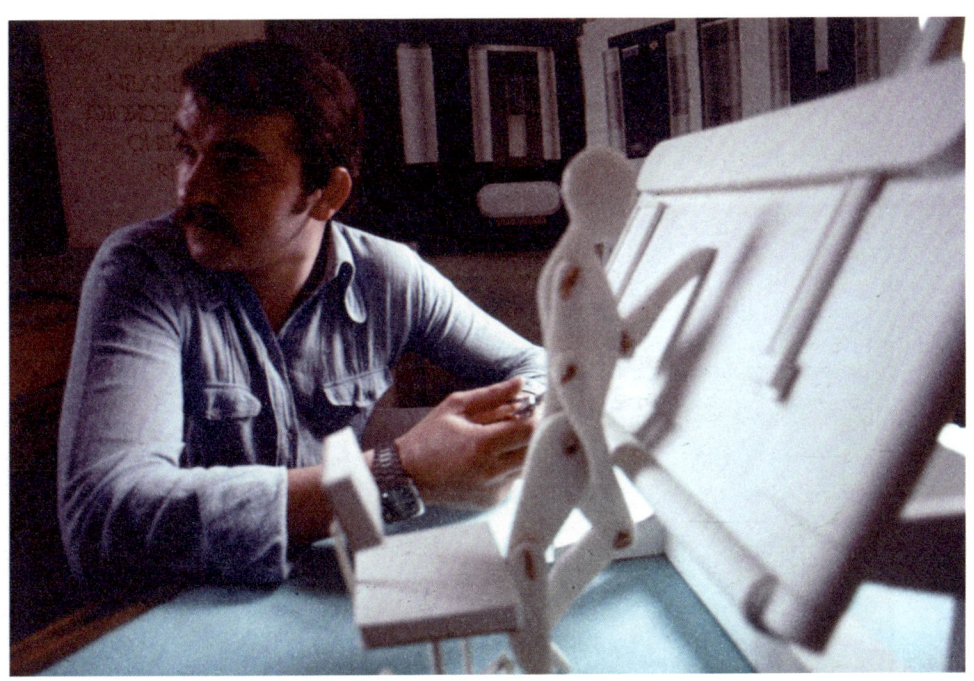

Domestic kitchen design featuring sectional
blocks with pull-out components

More than two million apartments came into use each year in the USSR and furniture designers were given the objective of satisfying the growing demand for kitchen furniture, electrical and gas equipment. VNIITE designers developed variations of kitchen furniture sets that consisted of modular sectional blocks. Each block was composed of areas to store food, diningware, appliances and various pull-out components. The sets were used in kitchens with different areas and layouts. With time, the variety of models was expanded and the designers were able to incorporate new technological advances in the manufacturing process and respond more directly to individual residents' needs. For example, at the start of the 1980s the designer Tatiana Suslova presented ensemble models called Dasha-1 and Dasha-2 that were developed to consider the needs of the elderly and the disabled (the project was awarded a prize at the Furniture-1983 competition). The variations of kitchen furniture developed by the VNIITE and introduced into mass production are shown on these pages.

Many of the VNIITE's designs for kitchens were
mass produced (see over)

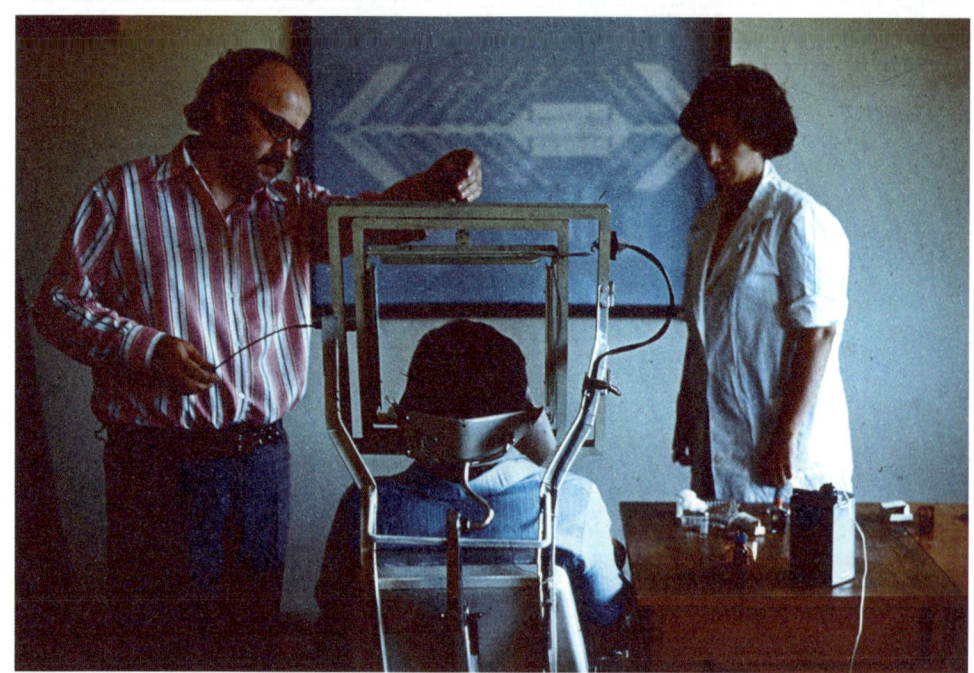

The Interdesign-80 seminar took place in Tbilisi,
Georgia in 1980 and explored designs for urban
living environments

Interdesign-80 was a project seminar organised by the International Council of Societies of Industrial Design (ICSID) in Tbilisi (Georgia) in 1980. Its main goal was to create a fully-fledged living environment that could be used in new areas of housing construction. Solving this challenge required an integrated approach and interaction between specialists from different backgrounds. Designers, architects, artists and graphic artists from 14 countries were invited to participate in the seminar, alongside the VNIITE and its branches, the USSR Architects' Union and the USSR Artists' Union. The French designer and architect, Pierre Vago, was invited as a moderator. The starting point was a detailed analysis of the construction of a new residential area in Tbilisi (it even included a colour scheme). At that time, integrated urban planning was only in its infancy and the seminar's work was its first serious test. The concepts presented by participants proved the effectiveness of a design-oriented approach to urban planning, while the ideas developed were also well received.

Photographs and drawings from the
Interdesign-80 seminar

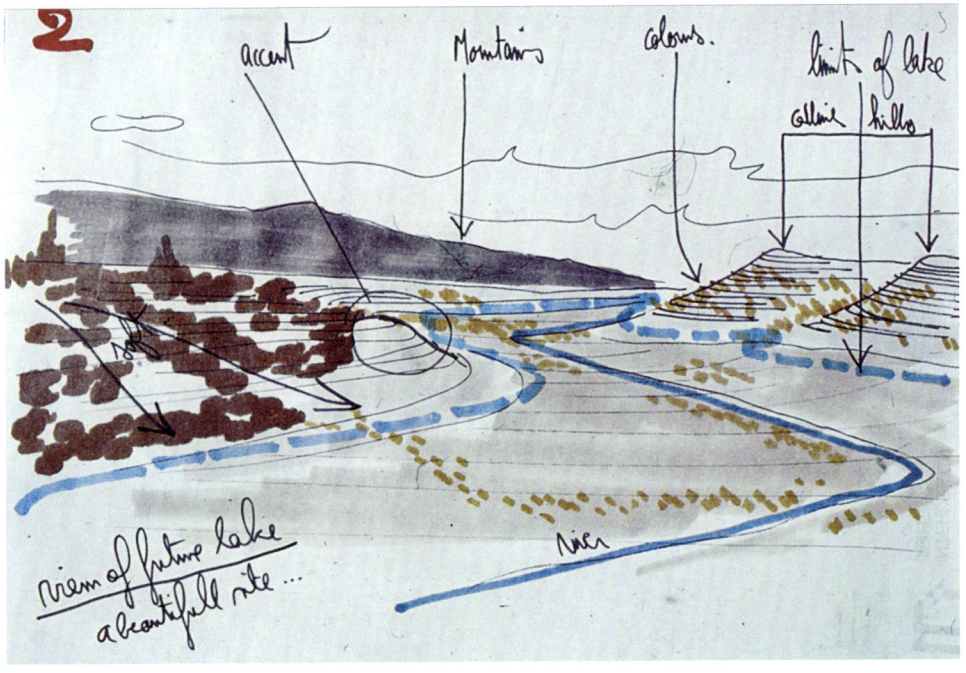

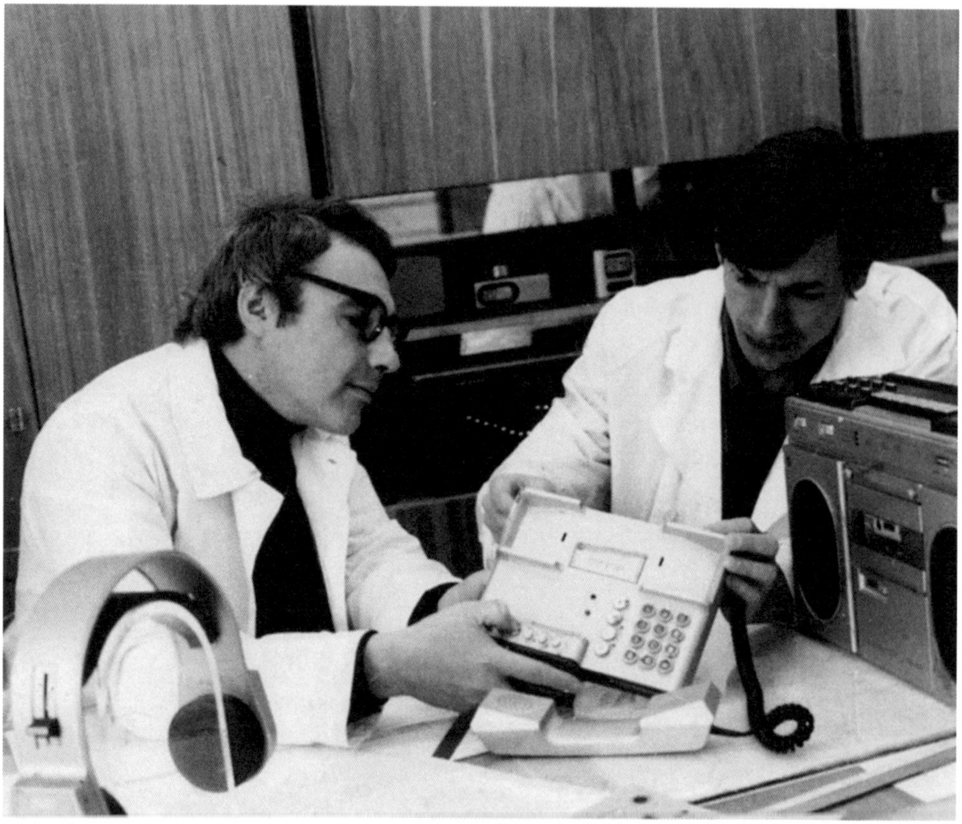

Designers working on various tape recorders for the
VNIITE's BAMZ programme

The following pages show a range of electronic equipment developed within the remit of the BAMZ ('Household tape-recording devices') programme. A large team of specialists worked on this important project from 1982 to 1986, including industrial designers such as Dmitry Azrikan, Valery Gossen, Alexei Kolotushkin, Lev Kuzmichyov, Marina Mikheyeva and others. During this time the USSR produced ten tape recorder models that had similar functions and a uniform design. The project began with a serious consumer study that resulted in the identification of different functional requirements and determined four preferences for style ('Classic', 'Instrumental', 'Youth' and 'Travelling') and three options for constructing the hardware – fixed, portable and for use in an automobile. To develop the product range, a three-dimensional model was used for the first time that enabled the design of 16 samples whose application methods and stylistic solutions differed: voice recorders and children's tape players; all-weather radio and tape players for young people; automobile and portable equipment; high-end tape recorders for music lovers; and experimental devices. The BAMZ programme established many of the methodological materials that would go on to help the development of system-wide design projects.

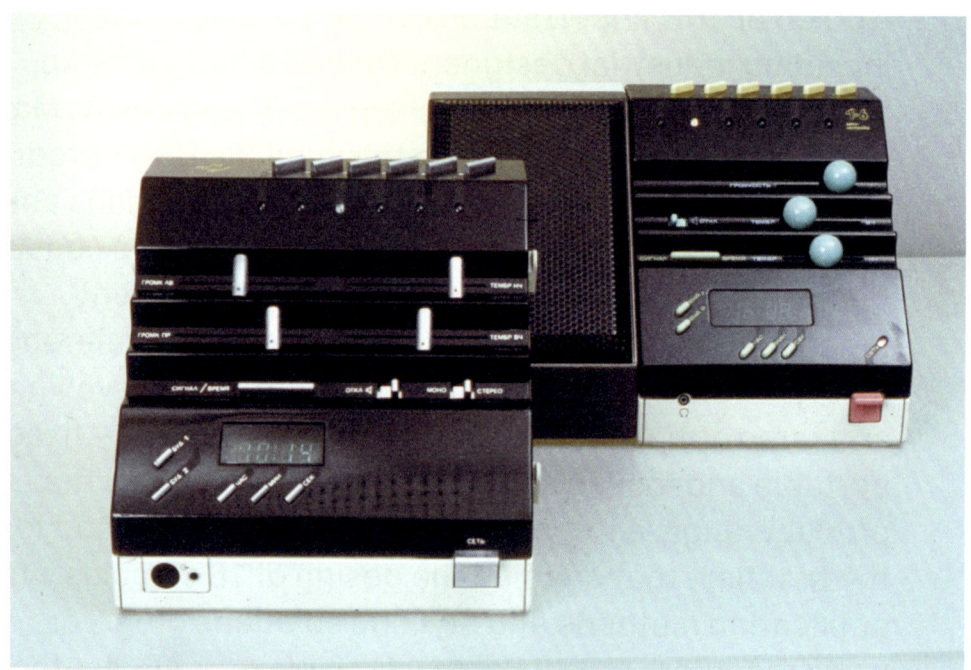

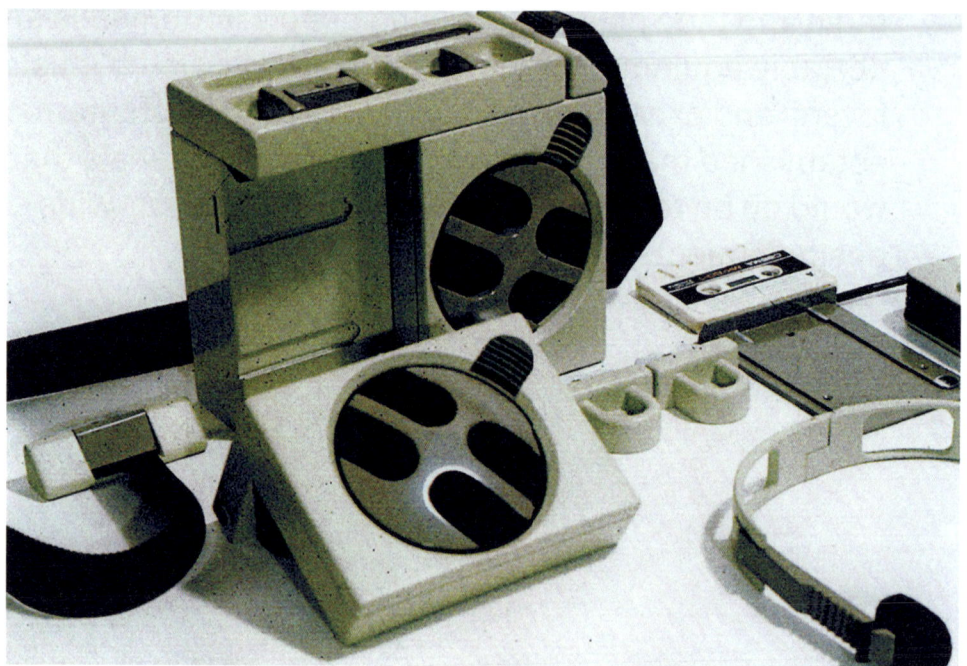

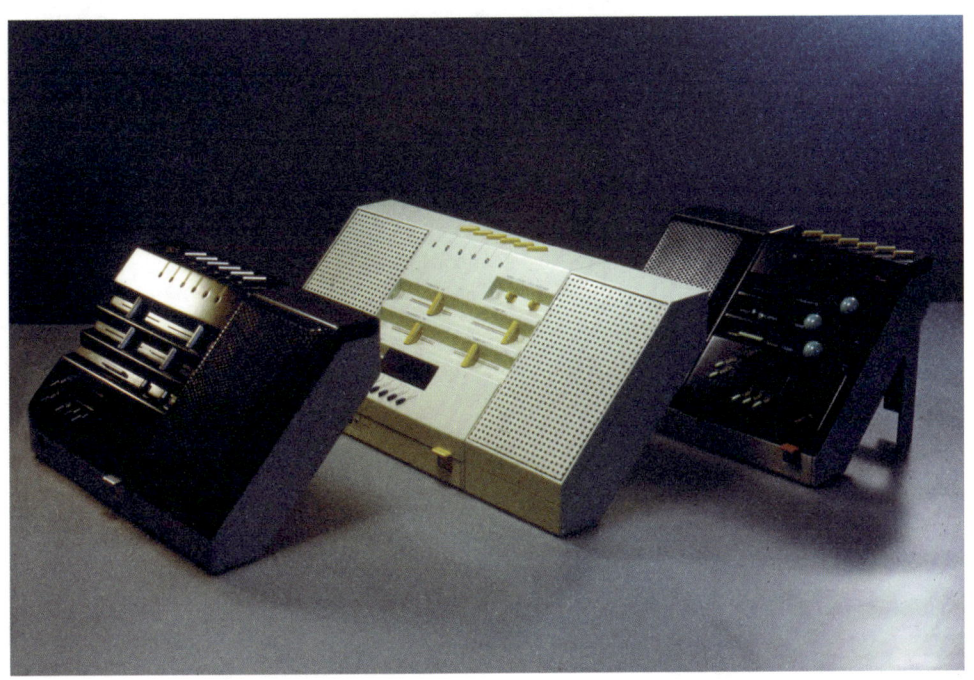

Electronic recording devices created by the
BAMZ design programme (see overpage).
The Saygak tape recorder is also featured on
page 162

The BAMZ programme focused on tape recorders
but also developed other audio devices

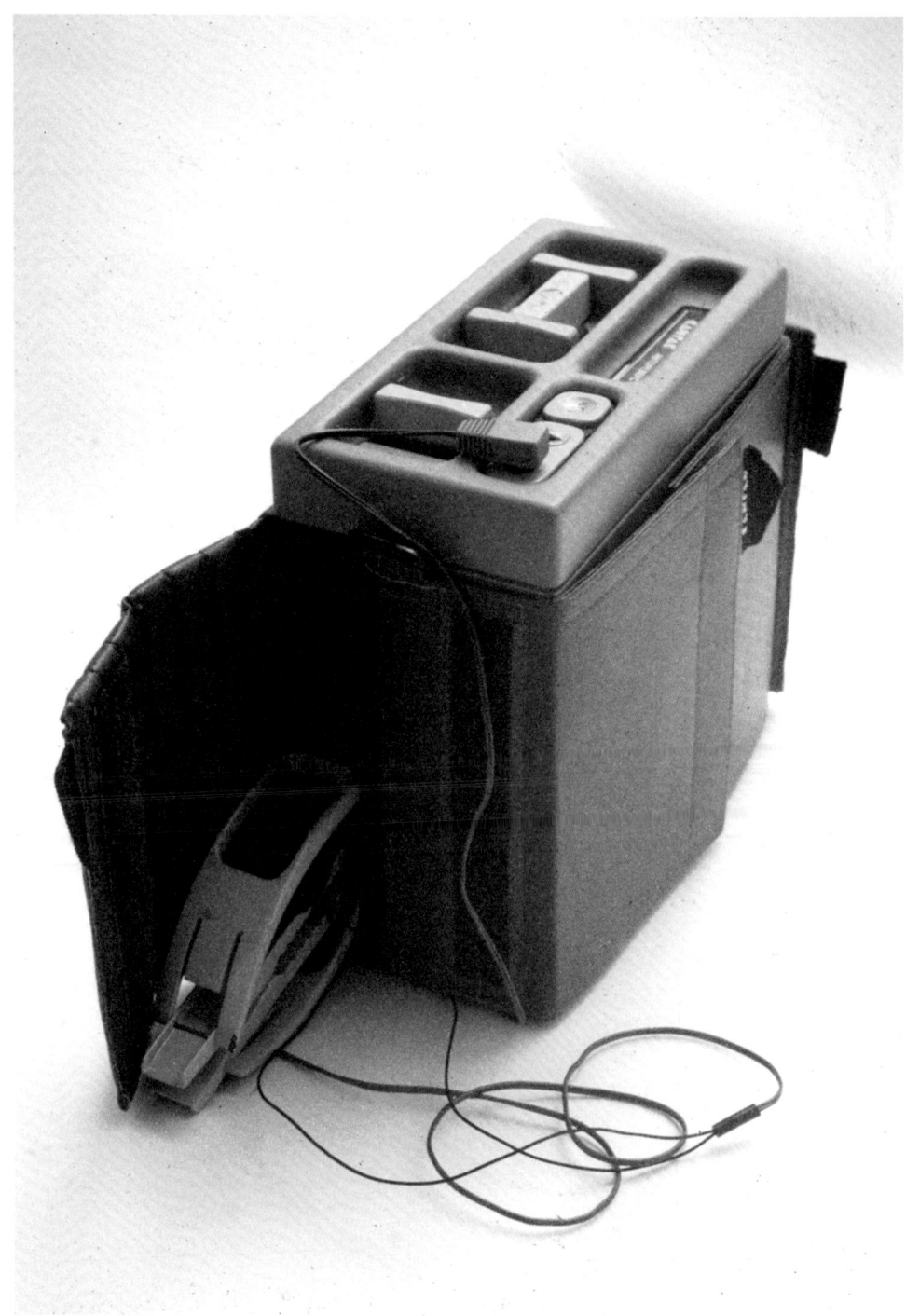

Portable Saygak tape recorder, designed in 1987 as
part of the BAMZ programme

From the mid-1980s, a trend arose for electronics to decrease in size and portable cassette decks began to appear. One of the landmark examples of this line was the universal Saygak tape recorder, developed by the VNIITE in 1987 as part of the BAMZ programme (see page 154), with the participation of industrial designers Dmitry Azrikan, Alexei Kolotushkin and Maria Kolotushkina. The designers faced the challenge of creating a portable but sufficiently sturdy player, since the radio and tape recorder was designed primarily for drivers and tourists. The Saygak fitted easily into a backpack and could even be attached to bicycle handlebars. The headphones remained large, since they were mostly manufactured as part of a fixed radio device (essentially a steel arc with miniature speakers attached). However, they were designed to fold away and had a special fastening (that resembled a clothespin) while there was also a compartment for carrying them in the case for the device. In terms of functionality, the Saygak was not inferior to many of the larger models, but the project was never executed and remained at the prototype stage.

The Saygak's carry strap and folding headphones
enabled the unit to be portable

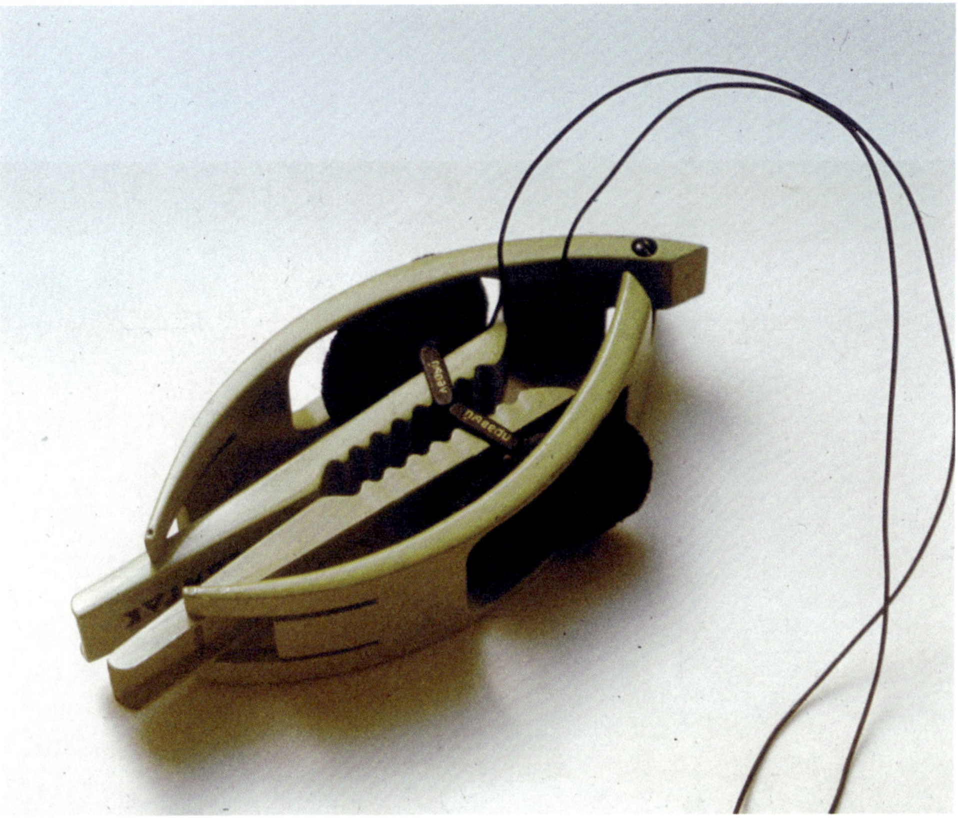

САЙГАК
АВТОРЕВЕРС
124
ТЕМБР
БАЛАНС

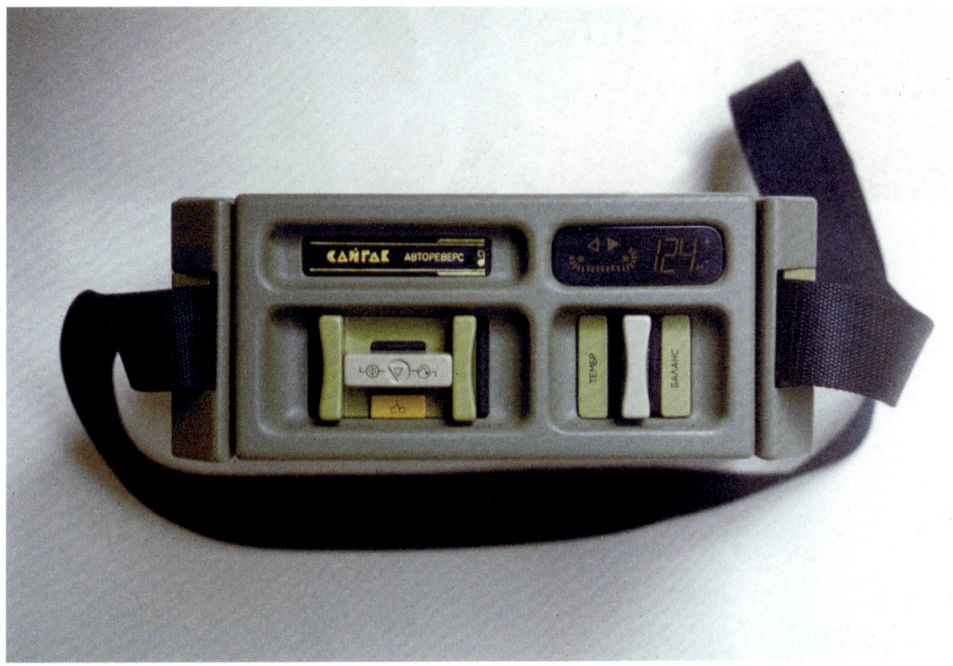

The Saygak tape player could even be installed
on a bicycle, as demonstrated by one of its
designers, Alexei Kolotushkin

Photographs showing some of the work created
on the experimental modelling seminars
(see overpage) held at the Centre of Industrial
Design Aesthetics

The following pages show the results of some of the scientific and practical seminars on experimental modelling conducted by the Department of Theory at the Centre for Industrial Design Aesthetics (CIDA) from 1984 to 1990. In total, seven cycles of classes were set up where practicing designers participated, as well as members of industrial design offices and scientific research institutes. The seminar leaders, Vyacheslav Koleychuk and Alexander Lavrentyev, selected a different area of design each time, holding a series of lectures and conducting practical classes on creating abstract, geometric compositions. For the organisers, it was important to introduce practicing designers to some of the era-defining creativity of the second half of the 20th-century, to teach them how to work with form as the basis of figurative and artistic communication, and to talk about the connection between art, mathematics, science and design. In this way, programmes on subjects such as 'Materials, technologies, and structure'; 'Visual culture – visual thinking'; 'Kinetic and dynamic forms in design'; 'Style'; 'The computer and visual culture in design'; and many others were established.

The Designing for a Socialist Society exhibition
of VNIITE projects (see also overpage) took place
in Moscow in 1985

The VNIITE organised more than 20 design exhibitions for the Council for Mutual Economic Assistance (COMECON) member countries. The most ambitious of them was called 'Designing for a Socialist Society' and was held at the VDNKh exhibitions centre in Moscow in 1985. It became a report on the development of design in the Soviet Union and other countries in the Socialist camp. While the main portion of the exhibition was dedicated to Soviet design, some of the other countries that participated included Poland, East Germany, Hungary and Bulgaria with real-life samples of equipment, industrial goods and interior design displayed. The event highlighted some of the most effective ways for long-term collaboration between designers and industrial workers and addressed the role design could play in the creation of a comfortable environment. There were also special displays for the achievements of each country that participated and a symposium devoted to the methods used in artistic design, the role of artistic traditions and experimentation in design and the interconnectedness of the latter with modern-day manufacturing and culture. More than 30,000 people attended the exhibition at VDNKh and similar exhibitions were held in other Socialist countries including Czechoslovakia, East Germany and Poland.

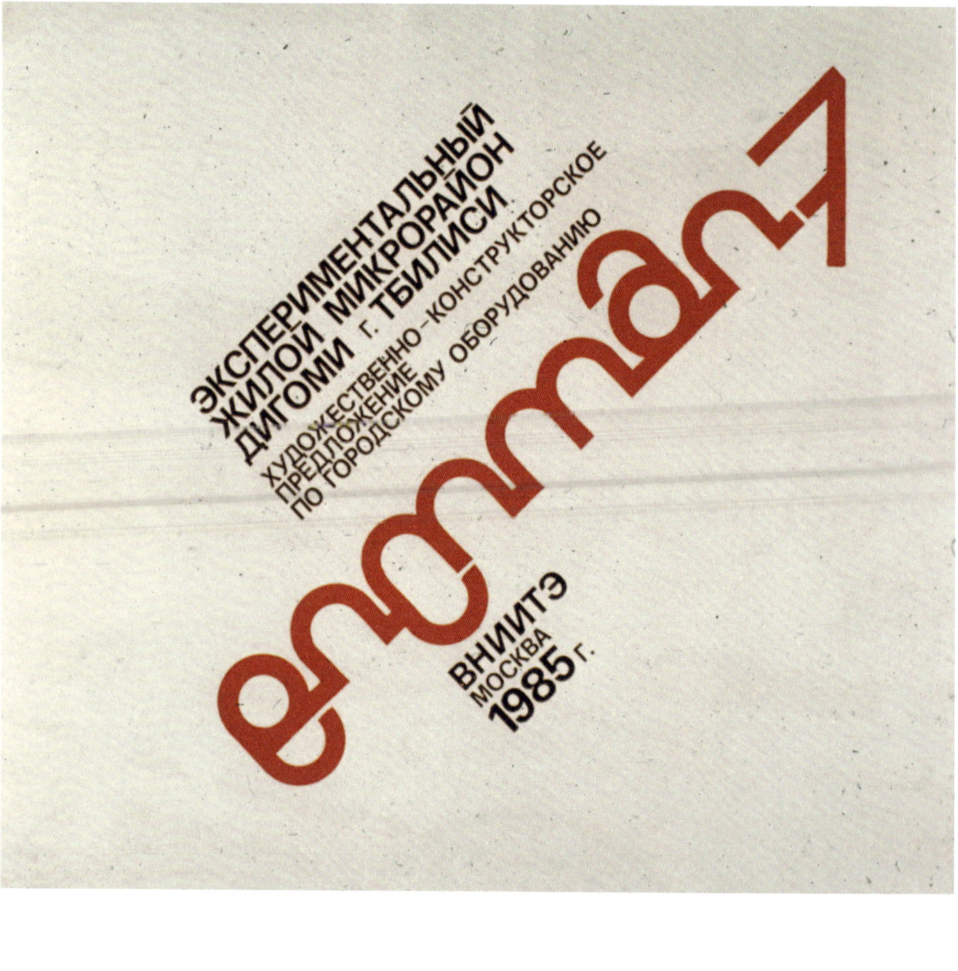

ЭКСПЕРИМЕНТАЛЬНЫЙ
ЖИЛОЙ МИКРОРАЙОН
ДИГОМИ г. ТБИЛИСИ

ХУДОЖЕСТВЕННО-КОНСТРУКТОРСКОЕ
ПРЕДЛОЖЕНИЕ
ПО ГОРОДСКОМУ ОБОРУДОВАНИЮ

ВНИИТЭ
МОСКВА
1985 г.

Cover for an instructional booklet created
for a series of urban installations designed
by the VNIITE

The following pages show the results of an urban planning project developed by the VNIITE in Digomi-7, a residential neighbourhood of Tbilisi, Georgia. Essentially an experiment in planning, architecture and design, the intention was to improve the functional and aesthetic qualities of the area and reflect the latest developments in the housing industry, as well as national and cultural traditions. In 1985, a group of industrial designers under the leadership of Dmitry Azrikan was given the task of developing a series of municipal installations. The team created five yellow modular units from which a series of different street installations could be assembled – from kiosks and water fountains, to benches and children's sandboxes. The project, which used inexpensive, vandal-resistant materials, was the first time that VNIITE architects and designers had worked together. Details of the project's development and implementation were published in the VNIITE's Technical Aesthetics journal.

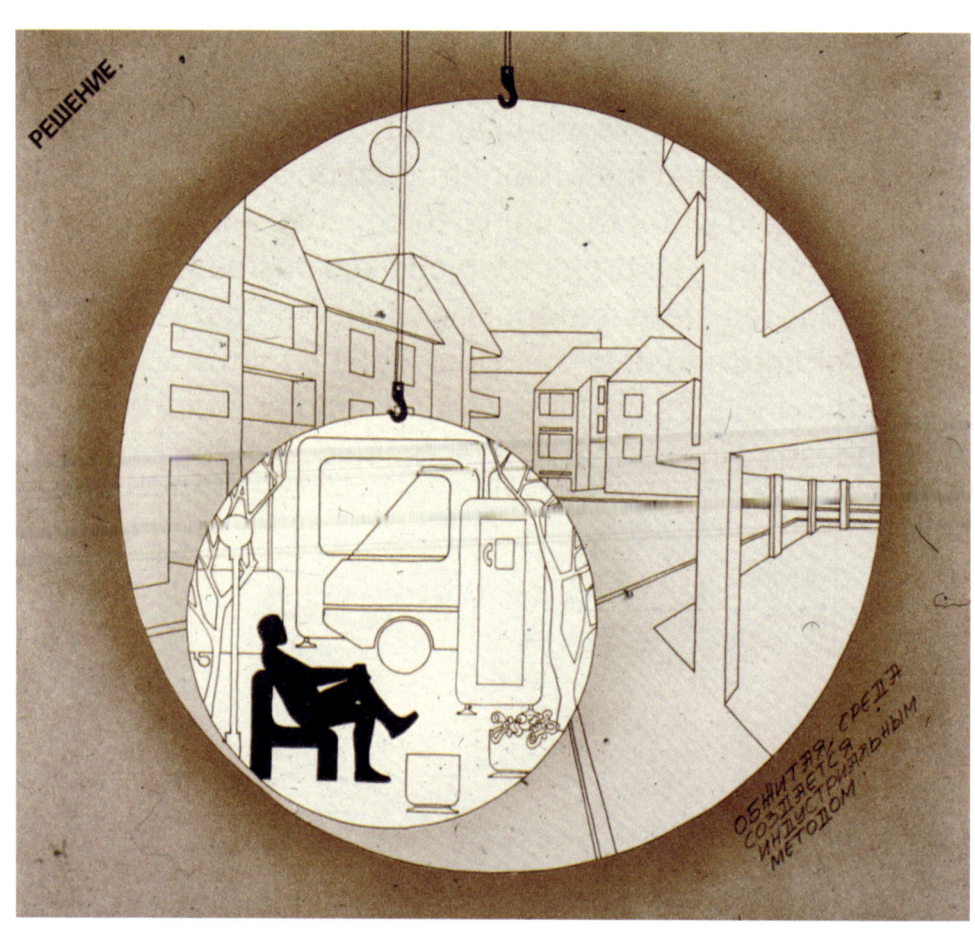

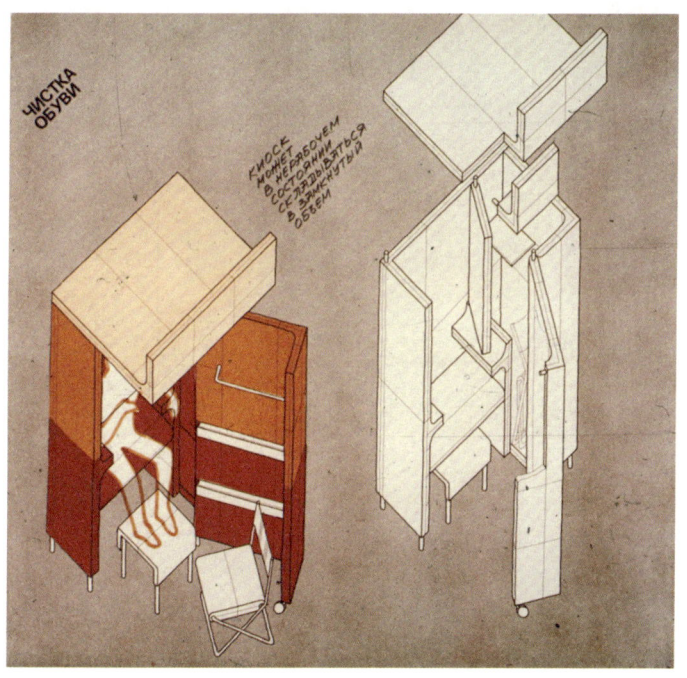

Pages from instructional booklet for VNIITE urban
installations, showing the various permutations of
the modular units

КИОСК

Pages from instructional booklet for VNIITE urban
installations

The modular units installed on the streets

Overpage, VNIITE designers Maria Kolotushkina
(on left) and Yelena Ruzova test the equipment
in a park setting

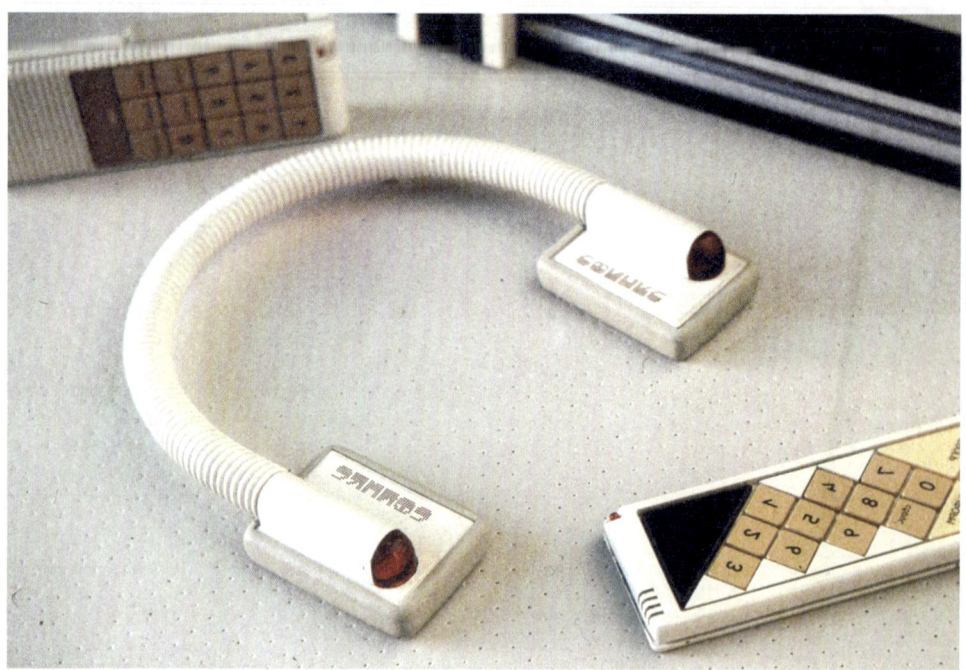

The Sphinx home studio system consisted of a series
of recording, storage and transmission devices,
complete with video screen and speakers (overpage)

The following pages show elements from one of the VNIITE's most famous projects – the SPHINX home radio and television studio. The name derived from 'Super-Functional Information Technology and Communications Complex' and it was designed in 1986 by a Dmitry Azrikan-led team of industrial designers including Igor Lysenko, Marina Mikheyeva, Yelena Ruzova, Alexei Kolotushkin, Maria Kolotushkina and others. The long-term goal was to equip all residences with these kinds of electronic devices by the year 2000 – and the project was focused on technical solutions that were, at the time, only at the experimental research stage. All work involving the receiving, recording and transmission of information was supposed to be accomplished using a central processor with a universal storage device and distributed throughout various rooms in an apartment on several LCD screens and via column loudspeakers. The purpose of these devices was wide-ranging – they not only encouraged family relaxation and game-playing, but also facilitated individual work and the management of kitchen devices and so on. The design proposal integrated the possibilities of the future into a single system but, unfortunately, the project remained at the prototype stage.

VNIITE designer consulting various designs created
for a new train for the Moscow Metro system

In 1987, the USSR Ministry of Automobile Industry announced the first nationwide design contest to develop a project for new trains for the Moscow Metro. Participants were asked to find a functional solution for the cabs and cars and increase the comfort of the carriage compartments as much as possible, given the amount of space assigned. One of the finalists in the contest was the 'Serebryanaya Nit' project, worked on by VNIITE industrial designers Lev Kuzmichyov, Dmitry Azrikan, Alexei Kolotushkin, Igor Lysenko, Marina Mikheyeva, and others. Rather than relying on one type of universal car design, a system was proposed with four modifications that specialised in running on different lines (and operated in different ways) – 'Line', 'Radius', 'Span', and 'Peak'. The assembly of each of the types of cars was to be accomplished using unified sections with a single universal cab – the differences were in the number of doors, the availability of space for baggage and the location of the seats. In addition to answering the requirements of the contest, the designers worked out an ergonomic workspace for the train driver, and they solved the problem of how to safely evacuate passengers in the event of an emergency.

Rendering and prototype of the Rapan caravan
designed by the VNIITE in 1988 (model shown,
overpage)

In 1988, a group of VNIITE industrial designers – Dmitry Azrikan, Alexei Kolotushkin, Marina Mikheyeva, Igor Lysenko, and others – designed a trailer on wheels called the Rapan. Residential trailers, so-called caravans, were very popular abroad, but in the Soviet Union this was the first development of its kind. The client was a subdivision of the Tupolev design office and this may, in part, account for the caravan's futuristic look. The Rapan was designed to be transported by automobiles of all types at a maximum speed of 80 kmph. Within an area of six square metres (and an inner compartment length of just over three metres), the designers managed to create a versatile living environment, which included a double bed for adults that transformed into a dining area, a two-storey bunkbed for children that turned into a play area, a full kitchen unit with a work surface, gas stove, small refrigerator, sink and draining rack, a wardrobe, a composting toilet and a shower cabin. The body of the caravan was manufactured from insulated panels and the possibility of extending it by adding modular sections was designed-in, as was a fixed plumbing unit. The project was transferred to the client and got approval, but was never realised.

Prototype Rapan caravan in action

VLADIMIR ARYAMOV (1924–1985)
Soviet designer and automobile engineer, one of the creators of the Soviet Union's experimental rear-engine cars. After graduating from the Moscow Automotive Institute, he was accepted by the Scientific Research Automobile and Engine Institute in 1953 on the recommendation of his teacher, Yuri Dolmatovsky. Aryamov initiated the use of soft upholstery, mouldings, radiator badges and other decorative elements on mass-production cars, designed vans, minivans, ambulances and off-road trucks. In 1963, he followed Dolmatovsky to the VNIITE, taking up a position of project architect in the Wheeled Land Transport Department.

DMITRY AZRIKAN (1934–)
Soviet, Russian and American designer, holds 106 Soviet/Russian and six American patents and certificates for industrial designs. He worked at the Yaroslavl Automobile Plant, then at the Neftekhimpribor Special Design Bureau, where he was Head of Design, before changing to the Special Design and Technology Bureau of Vehicle Fuelling Equipment. Meanwhile, Azrikan completed his PhD thesis via VNIITE postgraduate courses and began working at the institute in 1973, where he became Head of Future Design and Design Programme Development. In 1988, he founded his own studio and moved to the USA in the early 1990s where he became an Associate Professor and Head of Design at Western Michigan University and received accreditation from the US National Association of Schools of Art and Design. In 1998, he was awarded a Diploma of Honor of the Industrial Designers Society of America for dedicated service in training the next generation of IDSA leaders.

VALERY BERDYUGIN (1938–2006)
Architect, designer and photographer, assistant at ICSID. Chief Designer at the VNIITE's Ural branch, where he had been developing tramcars, high-speed trains, truck cabins, clocks and dynamometers. Berdyugin wrote for various VNIITE publications and worked for the publishing arm of the Russian Academy of Sciences's Institute of the History of Material Culture.

YURI DOLMATOVSKY (1913–1999)
Founder of the Soviet school of automobile design, member of the Board of Design of the Soviet Ministry for the Auto Industry, Board of Studies at the Faculty of Modern Engineering at the People's University of Moscow Motorists, Head of Automobiles at the Soviet Academy of Science's Central House of Scientists. His career began at the Automobile and Tractor Research Institute in the 1930s. Later, he moved to the Urban Traffic Research Institute, and then to the Automobile and Automotive Engine Research Institute. During WWII, he designed bus bodies, exterior truck components, and ambulances at the Stalin Automobile Plant No. 1 (later ZIL). After the war, he returned to NAMI as Head of the Body Design Bureau where he worked on modernising

Pobedas and Volgas and developed rear-engine cars. In 1963, he became Head of Wheeled Land Transport at the VNIITE, and continued as Head of the Laboratory of New Types of Motive Power at the Automobile Research Institute from the early 1970s. Dolmatovsky dedicated a lot of time to lecturing, writing and illustrating.

ALEXEY FEDOTOV (1938–)
Chief Designer of the Kirov Ust-Katav wagon-building plant SDB until 1982, developer of the most widely used tramcar in the world (KTM-5M3, 1971). Under his leadership, the Ust-Katav SDB developed a variety of tramcar models together with the All-Union Scientific Research Institute of Railway Car Building designers.

VALERY GOSSEN (1953–)
Soviet and Russian industrial designer and entrepreneur now living in Germany.

ALEXANDER GRASHIN (1936–)
Soviet and Russian designer and design theorist, full member of the Russian Academy of Natural Sciences and the European Academy of Natural Sciences, Honoured Worker of Science and Technology of the Russian Federation, holds around 50 certificates for industrial designs and projects. A painter, he later graduated from the Faculty of Mechanics at the Moscow Institute of Technology and obtained a post-graduate qualification at the VNIITE. Grashin headed the institute's Department of Design and Planning Theory, Methodology and Ergonomics. He has written numerous research works, introducing the 'design kit' and 'assembly kit' concepts, and was awarded many medals and diplomas, including a badge of honour from the International Union of Designers, and a gold medal from the Russian Academy of Arts.

RAMIZ GUSEYNOV (1940–)
Soviet and American graphic designer, member of the International Type Directors Club. He joined the VNIITE in 1971 after working under Dmitry Azrikan at the Neftekhimpribor SDB. The font he created for the institute's Elektromera design program became a National Standard. In 1984, Guseynov opened his own studio, specialising in corporate identity for major enterprises. In 1991, Guseynov moved to the US. Since 2006 he has run the TipografiaRamis foundry in Chicago.

ANDREI IKONNIKOV (1926–2001)
Soviet and Russian architect and historian, corresponding member of the Soviet Academy of Arts, Director (1966-1974), later Deputy Director (1980-1992) of the Architecture and Construction Engineering Theory and History Research Institute in Moscow, Head of Architecture at the Moscow Institute of Land Use Planning, and a Sector Head at the VNIITE (1979-1988). He co-founded the Russian Academy of Architecture and Construction Sciences, and was its Vice-President from 1994.

Ikonnikov was awarded a State Prize of the USSR in 1979, and a State Prize of the Russian Federation in 1992.

NIKITA KAPTELIN (1941–)
Soviet and Russian designer and artist. Kaptelin worked at the VNIITE from 1968, specialising in vehicle design, holds 31 certificates for industrial designs, more than ten certificates from VNIITE design competitions, several medals and diplomas.

SELIM KHAN-MAGOMEDOV (1928–2011)
Foremost expert on the history of the Russian avant-garde, Honoured Architect of the Russian Federation, Honoured Academic of Dagestan, honorary member of the Russian Academy of Arts, member of the Russian Academy of Architecture and Construction Sciences and the Moscow branch of the International Academy of Architecture. From 1978 to 1988, he ran a seminar on artistic issues of the substantial and spatial environment at the VNIITE's Department of Design Theory and History.

VALENTIN KOBYLINSKY (1931–)
Soviet and Russian designer, designed cafés and retail facilities before changing to the Automobile and Automotive Engine Research Institute under Yuri Dolmatovsky in 1955 where he designed vans, military vehicles and popular tractors. After working for the Minsk Automobile Plant, he began his best-known project – the development of the BelAZ-540 mining dump truck (1961), followed by numerous other heavy vehicles. When the VNIITE was founded, Kobylinsky became one of its first employees in 1962, working again under Dolmatovsky.

VYACHESLAV KOLEYCHUK (1941–2018)
Soviet and Russian artist, architect and designer, member of the UNESCO International Federation of Arts. After working at the bionic architecture laboratory at the Central Architecture Theory and History Research Institute until 1977, he was appointed Head of the Form Shaping group at the VNIITE's Department of Design Theory and History, a position he held until 1994. Having six inventions and more than 40 academic publications to his name, he is widely known as the founder of the kinetic art movement.

ALEXEI KOLOTUSHKIN (1956–)
Soviet and Russian designer, he joined the VNIITE in 1984 as a designer in the Department of Design Techniques where he worked on the BAMZ and SPHINX design programmes and visual identity. In 1988, he became Chief Designer at the Azrikan Studio, developing concept cars. Later, he headed design studios at Intertap Design and Dekos.

MARIA KOLOTUSHKINA (1960–)
Russian businesswoman and former designer. Until the 1990s, she worked at the VNIITE as a Future Design technician in the Department of Design Techniques under Dmitry Azrikan and contributed to the development of the BAMZ and SPHINX design programs.

IRINA KOSTENKO (1936–)
Soviet and Russian designer. After working at the Soviet Academy of Science's Institute of Organic Chemistry SDB and the Control Computer Research Institute, she joined the VNIITE in 1975 in the position of Lead Designer. She became one of the founders, and later Head, of the Centre of Technical Aesthetics. Following the break-up of the USSR, she has continued to work in design.

LEV KUZMICHYOV (1937–2015)
Soviet and Russian designer, member of the Russian National Academy of Design, the VNIITE's Director from 1987 to 2003, held six foreign patents and 90 certificates for industrial designs and inventions. After graduation from the Moscow Aviation Institute and the Moscow Higher School of Art and Design, he worked at the VNIITE, initially as Chief Designer, and then as Head of Industrial Design. Kuzmichyov was one of the founding fathers of the Soviet systematic approach to design, and of the development of the design programming method in the design of integrated systems. He was awarded a badge of honour, a Soviet Council of Ministers prize, and a Russian State prize.

ALEXANDER LAVRENTYEV (1954–)
Soviet and Russian designer, historian of design and photography, curator, Vice Rector for Research at the Stroganov Moscow State Academy of Arts and Industry. Lavrentyev joined the VNIITE in 1976 and worked in the Department of Design Theory and History under Khan-Magomedov and Koleychuk. As well as researching the history of Soviet design, he was a member of the VNIITE's editorial team, awarded a silver medal from the Soviet Academy of Arts in 1978, and later a certificate from the Academy of Graphic Design in 1998 and a Russian Government Prize for education in 2010. He is the author of more than 150 publications, among them works about his grandparents, Russian avant-garde artists Alexander Rodchenko and Varvara Stepanova.

IGOR LYSENKO (1961–)
Soviet and Russian designer and educator, holds 13 certificates for industrial designs. After graduating from VNIITE postgraduate courses, he worked there in systems research and design programmes under Dmitry Azrikan and was a Head of Sector until 1990, producing a number of large-scale designs for the electronics industry and medicine. After Perestroika, he worked as a practising designer for a number of large manufacturers (Panasonic, SONY, etc.) and started teaching.

ANDREI MESHCHANINOV (1940–2013)
Soviet and Russian industrial designer, Honoured Artist of the Russian Federation, honorary senator

of the Russian Union of Designers, held 26 certificates for industrial designs and patents. From 1968 to 1989, he worked at VNIITE's Leningrad branch in positions ranging from designer to Head of Integrated Research and Development. He was one of the organisers of the founding congress of the Soviet Union of Designers, and was on its Management Board from 1987 to 1991. From 1989, he ran a private studio. In 2004, he was awarded the badge of honour 'For Services to the Development of Design'.

MARINA MIKHEYEVA (1949–)

Soviet and Russian designer, member of the International Union of Designers, holds 30 certificates for industrial designs. Mikheyeva worked at the VNIITE from 1974 to 1988, rising from engineer to Chief Designer in Systems Research and Design Programmes, and continued her career at an independent design studio.

SVETLANA MIRZOYAN (1936–)

Soviet and Russian designer, design theorist and educator, holds 42 certificates for industrial designs. In 1962, Mirzoyan moved to Latvia and joined the VNIITE's Riga branch, where she stayed until returning to Leningrad in 1983. She later became the USSR's leading specialist on apron equipment design, was awarded a Certificate of Merit by the Russian Ministry of Education and Science (2005) and entered into the Book of Honour of the Saint Petersburg Union of Designers.

VLADIMIR MUNIPOV (1931–2012)

Soviet and Russian academic, founder of Soviet ergonomics, Honoured Cultural Worker of the Russian Federation, member of the Russian Academy of Education, International Commission on the Human Aspects of Computing, board of the International Foundation for Production Research, and Management Board of the Soviet (later Russian) Union of Designers. Munipov was one of the co-founders of the VNIITE, where he worked until 1992 and set up the country's first department of ergonomics. In 1972, he headed COMECON's newly created International Ergonomics Coordination Centre.

ALEXANDER OLSHANETSKY (1931–)

Soviet and Russian designer, board member and Vice-President of the International Union of Designers, holds 25 Soviet certificates for industrial designs and ten foreign patents. Having started his career as a designer at the Ministry for Shipbuilding's Central Design Bureau, he worked at the ZIL automobile plant and later at the VNIITE (1962-1987), helping to prepare the 'Emergency and special service vehicles', 'Vehicle number plates', and 'Aircraft maintenance and flight support equipment' national standards. He received numerous awards and later joined administrative staff of the Management Board of the Soviet Union of Designers.

ALEXANDER POPOV (1948–)

Soviet and Russian designer, holds 27 certificates for industrial designs. After graduating in 1972, he was allocated to the VNIITE, where he went on to become Chief Designer.

VLADIMIR RUNGE (1937–)

Soviet and Russian designer and researcher, one of the founding fathers of Soviet optical-mechanical and optical-electronic product design, co-founder of the Soviet Union of Designers, Honoured Artist of the Russian Federation, full member of the Russian Academy of Natural Sciences, presidium member of the Department of Artistic and Industrial Design at the International Academy of Natural and Social Sciences, honorary member of the Russian Academy of Arts, Vice-President of the International Union of Designers, awarded numerous prizes, medals and other awards. Chief Photography and Film Equipment Designer at the Krasnogorsk Mechanical Plant CDB, he was Chief Researcher at the VNIITE from 2008 to 2013.

YELENA RUZOVA (1962–)

Soviet and Russian designer and educator, member of the international Union of Designers and the UNESCO International Association of Art, holds certificates for industrial designs. After graduating, she worked at the VNIITE in systems research and design programmes under Dmitry Azrikan and started teaching in 1996. In 2010, she became an Honoured Higher Vocational Education Worker of the Russian Federation and received the Victoria Russian national design prize in 2011.

TATIANA SAMOILOVA (1939–)

Soviet and Russian designer, holds 42 certificates for industrial designs. At the VNIITE, she specialised in designing household appliances, manicure sets, clocks, and hardware products for administrative and residential buildings.

DMITRY SHCHELKUNOV (1940–)

Soviet and Russian designer, member of the World Future Studies Federation, co-founder of the International Design for Extreme Environments Association, holds more than 50 certificates for industrial designs. Shchelkunov has worked at VNIITE since 1965, heading the Future Design workshop. He has produced design concepts for household and industrial appliances, power tools, machine tools, computing and communications equipment, medical apparatus and participated in many design programmes, including survival equipment for cosmonauts and the interior of the Mir-2 space station. He has developed 14 National Standards on technical aesthetics and ergonomics.

TATIANA SHEPELYOVA (1922–2010)

Soviet automobile designer working at the VNIITE until the 1970s, developing a number of prospective projects and experimental prototypes.

YURI SOLOVIEV (1920-2013)
Leading Soviet and Russian designer, founder and
Director of the VNIITE, founder and President of
the Soviet Union of Designers, a member of the
Congress of People's Deputies of the USSR, Deputy
Chair of the Culture Committee of the Supreme
Soviet of the USSR. In 1945, he established the
USSR's first design organisation specialising
in industrial design, the Architecture and Art
Bureau of the People's Commissariat of Medium
Mechanical Engineering. Soloviev headed the
VNIITE for 25 years, being simultaneously Editor-
in-Chief of Tekhnicheskaia estetika (Technical
Aesthetics). He served as Vice-President of ICSID
from 1969 to 1976, and as its President from
1977 to 1980, before becoming a member of
the Senate in 1981. Soloviev was awarded many
international prizes, a Russian Presidential Award
for Literature and the Arts in 1999, and medals for
his exceptional contribution to the development
of design education and for his services to the
development of design, both in 2001.

TATIANA SUSLOVA (1943–)
Soviet and Russian designer and educator, full
member of the Russian Academy of Quality Issues
and a member of the state attestation commission
at the Stroganov Moscow State Academy of Arts
and Industry. After gaining her first degree, she was
assigned to the VNIITE as a designer. There, she
earned a postgraduate qualification and became
a Lead Researcher, working on ergonomic design
for the disabled and the elderly and winning many
competition prizes.

ALEXANDER YERMOLAYEV (1941–)
Soviet and Russian architect and designer, honorary
member of the Russian Academy of Arts. Yermolayev
began working at the VNIITE in its early years. In
1999, he was awarded the Victoria Russian national
design prize and a prize from the Russian Academy
of Graphic Design. From 2000 to 2006, he served on
the expert panel for the Russian Presidential Design
Award. In 2002, Yermolayev's studio was recognised
by the Russian Union of Designers for its high
professional design standards.

STANISLAV ZARITSKY
Design engineer at the VNIITE's Ural branch in
the 1970s, he developed tramcars and truck
cabins. Zaritsky's Molnija-295-4BN clock (1967)
was regarded as one of the USRR's best
industrial products.

In 2016, Discovering Utopia: Lost Archives of Soviet Design was exhibited at the inaugural London Design Biennale, where it was awarded the 'Utopia' medal. The majority of the images featured in this book were first shown at this exhibition.

Curator:	Alexandra Sankova
Researcher and editor:	Olga Druzhinina
Project management:	Natalia Goldchteine, Ekaterina Shapkina
Exhibition design:	Stepan Lukyanov
Video:	Svetlana Chirkova

I first saw the Discovering Utopia exhibition while judging the first London Design Biennale in 2016. The theme was 'Utopia' and all methods of expressing it were welcome. Some entries adhered to the theme subliminally and some in a more expected manner.

The Russian team's response to the theme perfectly encapsulated it. By exhibiting photographs and models of realised and unrealised pieces of design from Communist Russia side-by-side, the exhibition demonstrated the modern, innovative quality of the design and exposed the dilemma inherent in the act of creating a utopia. It's impossible to realise perfection in planning for all because humans are imperfect beings.

Yet, rather than be discouraged by this, the exhibition asked viewers to balance the Communist desire for a perfectly-designed world against the real world of human competitiveness and inequality.

Paula Scher, Partner, Pentagram New York

COLOPHON

VNIITE
Discovering Utopia:
Lost Archives of Soviet Design

UNIT 40

Texts:
Alexandra Sankova
and Olga Druzhinina

Text Editor:
Tatiana Zborovskaya

Editor:
Mark Sinclair

Design:
spin.co.uk
Tony Brook, Claudia Klat

Design Assistants:
Tim Frei, Charlotte Viglino

Production Manager:
Edie Lippa

Publishing Director:
Patricia Finegan

Typeface:
PX Grotesk

Printer:
Verona Libri

ISBN: 978-1-9164573-0-0

post@uniteditions.com

© 2018 Unit Editions

THANKS

Academic consultants
to the publication

Alexander Grashin
Alexander Lavrentyev
Tatiana Matveyeva
Marina Timofeyeva

We are grateful to the following
people for their assistance in
preparing the publication

Lev Khabarov
Alexei Konoplev
Vitaly Mikheev
Tamara Olshanetskaya
Ilya Polyanskikh
Yulia Voronkova

Unit Editions would like to thank
the following

Alexandra Sankova, Olga Druzhinina and
Tatiana Zborovskaya from the Moscow
Design Museum for allowing us to make
this book. Thanks to Paula Scher for
drawing our attention to the archive
and for her written contribution, and to
Quentin Newark for his help in getting
this book off the ground.

All images from the Moscow Design
Museum Collection

МОСКОВСКИЙ
МУЗЕЙ ДИЗАЙНА
MOSCOW DESIGN
MUSEUM